LIBBY PRISON BREAKOUT

LIBBY PRISON
BREAKOUT

The Daring Escape
from the Notorious
Civil War Prison

JOSEPH WHEELAN

PublicAffairs
New York

Designed by Brent Wilcox
Text set in 11 point Janson Text

Library of Congress Cataloging-in-Publication Data
Wheelan, Joseph.
    Libby Prison breakout : the daring escape from the notorious Civil War prison /
Joseph Wheelan.—1st ed.
        p.   cm.
    Includes bibliographical references and index.
    ISBN 978-1-58648-716-4
    1. Libby Prison.   2. Prisoner-of-war escapes—Virginia—Richmond—History—
19th century.   3. Tunnels—Virginia—Richmond—History—19th century.
4. Prisoners of war—Virginia—Richmond—History—19th century.   5. Prisoners of
war—United States—History—19th century.   6. United States. Army—Officers—
History—19th century.   7. Escaped prisoners of war—United States—History—19th
century.   8. United States—History—Civil War, 1861–1865—Prisoners and prisons.
9. Richmond (Va.)—History—Civil War, 1861–1865.   I. Title.
    E612.L6.W48 2009
    973.7'72—dc22
Paperback ISBN 978-1-58648-908-3
E-book ISBN 978-0-7867-4627-9

For my wife, Pat, and our daughters, Sarah and Ann

CONTENTS

Prologue   ix

1 The Confederate Capital   1

2 The POW Archipelago   19

3 Inside Libby Prison   31

4 The Defiant Colonel   47

5 Misery and Retaliation   65

6 Miss Van Lew's Spy Ring   83

7 The Warrior Schoolteacher   103

8 Test of Faith   121

9 General Butler's Raid   129

10 The Ordeal of Tunnel Four   143

11 Escape   157

12 Flight   169

13 Fallout   193

    Epilogue   219

Acknowledgments   231

Notes   233

Bibliography   257

Index   273

PHOTO INSERT BETWEEN PAGES 142 AND 143

PROLOGUE

EARLY SUMMER, 1863

After enduring several weeks in Richmond's Libby Prison, 1st Maine cavalryman Edward Tobie was being released. He had been captured weeks earlier at Brandy Station, Virginia, during the war's biggest cavalry battle. Now he was benefiting from the prisoner exchange system created by the warring sides a year earlier. On this warm morning, the Confederates marched Tobie and more than a dozen other Union officers from the prison overlooking the James River to the Richmond railroad depot. There they crammed the officers and enlisted parolees into sweltering cattle cars. Then the train clattered over the river bridge and entered the green rolling hills stretching south of the Confederate capital.

After traveling twenty miles, the train stopped at Petersburg, and the prisoners were permitted to briefly leave the boxcars and buy food. When they returned, the train traveled northeast for another ten miles, finally stopping in City Point. Located at the confluence of the James and Appomattox rivers, City Point was a small but important river port established in the seventeenth century during Virginia's early colonial days.

Tobie and his comrades detrained and assembled at a wharf to await the "flag of truce" boat, which routinely plied the river between City Point and Union-occupied Fortress Monroe, located at the James's estuary in Chesapeake Bay. The boat represented a rare rapprochement between the enemies. It ferried paroled captives, mail, and boxes of foodstuffs from North to South, and from South to North.

The parolees sought the shade of the leafy hardwood trees along the river for relief from the oppressive Southern sun. As they waited, they cast withering looks at the "white-livered" Confederate flag barely stirring in the heavy air.

✳   ✳   ✳

The war was in its third year, and neither the Union nor the Confederacy had yet gained a decisive advantage. The Union had prevailed at Shiloh and Antietam, but the Rebels had triumphed twice at Manassas, as well as at Fredericksburg, and most recently, at Chancellorsville. Neither side could feel confident of achieving a culminating victory. Each successive battle brought a steeper butcher's bill and a flood of new war captives to the Richmond prisons.

Caring for the prisoners would have been burdensome for both sides had it not been for their agreement in July 1862 to systematically parole and exchange war prisoners. City Point was one of the two designated sites where paroled Union captives embarked for home, and where Confederates released in the North arrived in the South. The other exchange site was on the Mississippi River, near Vicksburg. The first prisoner exchange took place on August 3, 1862, on the James River, with the paroling of six thousand Union and Confederate captives.

Since that day, about two hundred thousand soldiers had been sent home from prisons in the North and South. When the cartel was operating smoothly, captives were held for relatively brief periods, and were, for the most part, decently housed and fed.

Although it was important to both the Union and Confederacy to recover their captured soldiers, the Rebels needed them more urgently than did the Yankees. The South's free population of 5.5 million—its 3.5 million slaves were ineligible for army service—was just one-fourth the North's 22 million residents, nearly all of whom were free. Moreover, it was a struggle for the Confederacy to provide for the prisoners, and so they were anxious to parole the Yankees as quickly as possible. The Union, with its vaster resources, had no problem feeding and caring for its Confederate captives.

Even in the spring of 1863, when the cartel was functioning well, Confederate officials, beset by shortages, had triaged food distribution, ranking the needs of the war captives well below providing for the Confederate Army and Southern civilians. "I would rather they should starve than our own people suffer," wrote the Confederate exchange commissioner, Colonel Robert Ould.

A plume of smoke appeared downriver, and the soldiers could hear a laboring engine and churning water. It was the flag of truce boat, sailing under a standard of stars and stripes. The Yankees cheered; tears streaked their grimy, bearded cheeks.

Before getting on the boat, where they would receive medical care and a meal of bread, boiled ham, and coffee, the parolees had to wait while the Confederate parolees debarked. Lieutenant Clay MacCauley of the 126th Pennsylvania, taken prisoner at Chancellorsville, noted a stark contrast between the returning Confederates and his fellow Yankees. The Rebel prisoners, he wrote, were "a well-fed and vigorous reenforcement for the armies of the rebellion [while] our government authorities [got] a famished, exhausted, crippled, and seriously injured body of men."

Even as the exchange cartel was sending Tobie and MacCauley home, it was breaking down. President Abraham Lincoln's Emancipation Proclamation and the Union Army's subsequent aggressive recruitment of thousands of black soldiers had struck a raw nerve in the Confederacy. The Confederate government declared that it would exchange neither captured black Union soldiers nor their white officers.

Because the Emancipation Proclamation had signaled the Union's determination to end slavery, the Lincoln administration would not accept the Confederacy's position. War Secretary Edwin Stanton was poised to announce the cartel's suspension.

As accusations flew between the adversaries, the Union captives in Richmond steeled themselves to pay for the governments' obstinacy.

1

The Confederate Capital

"Death held a carnival in our city. We lived in one immense hospital, and breathed the vapors of the charnel house."

**—Sally Putnam, describing Richmond
during the Seven Days battles of 1862**

SEEING RICHMOND for the first time in 1861, a smitten T. C. DeLeon wrote that the city "burst beautifully into view, spreading panorama-like over her swelling hills. . . . No city of the South has [a] grander or more picturesque approach."

Indeed, when the war began, the stately "City of the Seven Hills" was an island of peace and hope in the tempestuous South. One of the last holdouts against war, Virginia had tried to broker a peace on the very day that its sister states met to form the Confederate States of America. Its capital, Richmond, instead became the iconic symbol of a bloody, fratricidal war. No other nineteenth-century American city would experience so many giddy triumphs and crushing losses in so brief a period.

Built on amphitheater-like hills along a curve in the James River, 125 miles from the Atlantic Ocean, Richmond was the South's third largest city in 1861, with 38,000 people. It had succeeded Williamsburg as Virginia's capital city in 1780, just when America's first colony was becoming the main theater of the Revolutionary War. A year later, Richmond was burned by British troops—the same hard-handed treatment that British soldiers, thirty-three years later, visited on Washington, D.C., Richmond's Civil War doppelgänger.

1

Richmond was Virginia's chief inland port for oceangoing ships, averaging shipments of 100,000 tons per year, primarily cotton, tobacco, grain, and finished goods made in Virginia and Europe. Five railroads crossed paths in the city, and there were twelve flour mills, but tobacco was king; Richmond was home to fifty-two tobacco-makers, seven major tobacco warehouses, six tobacco-box factories, and two cigar-makers. "The atmosphere of Richmond is redolent of tobacco; the tints of the pavements are those of tobacco," observed a British visitor.

More consequential once hostilities began were Richmond's foundries; their quality was unrivaled anywhere south of the Potomac. Iron from the James River area was reputedly the best in America, and before the war, Richmond's iron-makers ranked among the nation's top defense contractors, producing cannons in great quantities for the U.S. Army and Navy. With the outbreak of war, these manufacturers and their cannons, arms, ammunition, and heavy equipment served the Confederacy alone. In Richmond's industrial district along the James were four rolling mills, fourteen foundries and machine shops, nail works, six places where iron railing was made, and fifty iron-and-metal shops. Coal mined from seams twelve miles from Richmond fed the forges. Richmond was the Confederacy's Ruhr and its Birmingham.

Looming over the scores of foundries, mills, and shops was the massive Tredegar Iron Works, with its slate roofs and tall, smoking chimneys. During the war, it produced about half the Confederacy's cannons. Operated by Jos. R. Anderson & Co., at the foot of Gamble's Hill and between the James River and the Kanawha Canal, Tredegar, named for its Welsh builder, employed nine hundred men—fully 60 percent of Richmond's iron industry workforce.

Before the war, Tredegar made locomotives; cannons; steel for rifles, pistols, and bayonets; steel for Virginia railroads; pots for North Carolina kitchens; and "filigree material" for Georgia blacksmiths. Tredegar-owned slaves, specially trained by Joseph Anderson to perform skilled jobs, made up two-thirds of the mill's pre–Civil War workforce—unusual in the agricultural South, where slaves usually worked as field hands and servants. Anderson also used the slaves to break a strike by white workers—who were protesting having to compete with the slaves.

Tredegar's workforce expanded during the war to two thousand slaves and free workers, and it became a pillar of the Confederate war effort. It manufactured the rifled 7-inch "Brooke gun"; torpedoes; plates for ironclads, including most of the *Merrimack*'s armor; and shot, shell, and fortification guns.

*   *   *

Quiet but charged with anticipation when the war began in April 1861, Richmond became the Confederate capital when Virginia invited the secessionist government to relocate from Montgomery, Alabama; Montgomery's remoteness from ongoing military operations had been a drawback. By July 20, the Rebel administration was up and running in Richmond. The city was never the same. Between 1861 and 1863, Richmond's population grew from 38,000 to more than 120,000. "The city was thoroughly jammed," wrote T. C. DeLeon, a South Carolina diarist who was an aide to President Jefferson Davis. Living quarters were difficult to find at any price, and hoteliers put extra beds in hallways and parlors to accommodate the overflow. After Judith McGuire fled with her family from their home in Alexandria, Virginia, never to return, she walked Richmond's streets for days, futilely inquiring at boardinghouses for a room. "I do not believe there is a vacant spot in the city," she wrote in her diary in disconsolation. Sally Putnam, a writer, diarist, die-hard Confederate patriot, and member of Virginia's gentility, described her home as "a city of refuge. . . . Flying before the face of the invader . . . the homeless and destitute crowded into our city for safety and support." The refugees rented lodgings that Putnam described as being of "India-rubber capacity," because of their uncanny ability to absorb seemingly limitless numbers of lodgers. And with practically every skilled worker and mechanic in the army, too few remained to put up new housing.

Laborers, retailers, and contractors from the Confederacy's eleven states poured into Richmond seeking jobs, profits, and government war contracts. At times, confessed Putnam, Richmond seemed to her "a strange city, from the signs over the doors of the shops." By this, she meant that many of the new businesses bore Jewish names: "Israel and David, and Moses and Jacobs, and Hyman and Levy, and Guggenheimer and Rosenheimer, and other names innumerable of the Ancient People."

Richmond was transformed as much psychologically as physically. No longer just the capital of Virginia, it was now the capital of the Confederacy. Already an important Southern commercial and industrial center, now it was the South's most vital center. Because it lay within striking distance of the Army of the Potomac, Richmond became the fulcrum of Union and Confederate military strategy, and the Confederacy's overriding security concern.

Being at the Rebellion's nexus, Richmond citizens almost overnight became fiercely loyal to the Confederate States of America—an extraordinary change in view of Virginia's initial reluctance to join the Confederacy. A visitor from France was impressed by the city's "warlike spirit that nothing, not even the fear of death could daunt." There was no tolerance of "the miserable croakers who ever look on the dark side of the picture," as Putnam described them, much less anyone sympathetic to the North. Still, many ardent Unionists chose to remain, while being careful to conceal their sympathies from their rabidly partisan neighbors.

✳ ✳ ✳

No sooner had the Confederate government begun settling in than Rebel troops defeated Union forces in the first major battle of the war, at Manassas. In a striking gesture of unity, much of Richmond turned out to meet the trains from Manassas filled with wounded Confederate soldiers. "All strained forward in the same intent gaze, as car after car was emptied of its ghastly freight," wrote T. C. DeLeon. At a mass meeting called by Mayor Joseph Mayo, Richmond residents eagerly donated $8,000 for the care of the 1,336 wounded men. The first bandaged soldiers returning to Richmond were objects of curiosity and pity. But as the battles grew in size and destructiveness, men on crutches, many missing arms and legs, became a common sight on the city's streets.

In February 1862 Jefferson Davis became the Confederacy's "permanent" president in a ceremony in Capitol Square held before a large, cheering crowd in a pouring rain. A week later, he imposed martial law on Richmond. This was a response to numerous petitions by citizens upset by the spiraling disorder caused by the drifters, speculators, and criminals who had converged on the capital. Reluctant to trample on civil liberties, Davis and the Confederate Congress at first had tried milder measures, such as arming local police and imposing a curfew for alcohol sales. But when these steps failed to quell the lawlessness, Davis named a provost marshal for Richmond—sixty-two-year-old General John Winder, the inspector general of Richmond's military camps. Winder became the city's de facto dictator.

Jefferson Davis had respected Winder since their Mexican War service and Davis's days as U.S. secretary of war, and he valued Winder's loyalty and competence. The inspector general's duties had already expanded beyond their original scope of overseeing the city's military camps when he was named provost marshal; he also supervised Richmond's war prisons, investigated disloyalty charges, and issued passports to people traveling to the North.

Competence and a relentless quest for vindication of his family's name had taken Winder far in the U.S. Army—to general rank and a high position in the Department of Subsistence before the war. He bore the name of a proud Maryland family that for two centuries had produced prominent planters and soldiers. But in 1814, British troops routed the American army commanded at Bladensburg by Winder's father, General William Henry Winder. Later that day, the British marched into Washington and burned the Capitol, the President's House, and the Treasury. This searing event occurred during fourteen-year-old John Winder's plebe year at West Point. It confirmed for him his decision to make the army his career, and he spent the rest of his life striving to remove the blot from the family name. He became a capable officer, but one who was also known for being short-tempered and stubborn. During the Mexican War, where so many future Civil War generals underwent their baptism by fire, Winder distinguished himself at Contreras, Churubusco, and Chapultepec, and then served as lieutenant governor of Vera Cruz. Between wars, Winder made the South his home, marrying a North Carolinian widow, Caroline Cox Engles—she was Winder's second wife; his first wife had died—and buying a plantation near Wilmington, North Carolina. In 1861, he resigned his U.S. Army commission and joined the Confederate Army.

As provost marshal, Winder quickly grew unpopular for meddling in Richmond affairs. He broke up a lithographer strike and fixed newspaper prices at a nickel a copy because the city's newsboys had been overcharging their customers. In the interest of public safety, he cracked down on reckless army teamsters and carriage drivers and enforced a 5 mph speed limit. Necessary though these measures were, each offended a faction, and the diverse grievances produced a groundswell of resentment.

To flush out traitorous Southerners, Winder needed detectives. He consulted friends in Baltimore, a Union city under martial law because it was home to a large number of Confederate sympathizers and the main roads and railroads to the U.S. capital passed through it. His Baltimore contacts helped him assemble a detective force of thirty "plug-uglies," so nicknamed for their violence and hard drinking. Winder's "provost guards" closed saloons and distilleries and confiscated citizens' firearms and swords. They raided gambling dens and brothels. Their sting operations snared druggists, who, for a bribe, filled phony medicinal "prescriptions" for alcohol, though generally only after the plug-uglies had gulped down the prescriptions. The men were so widely hated and blatantly corrupt that Winder had to fire most of them within months.

* * *

RICHMOND, JUNE 1862

The thudding cannon fire east of the city reminded citizens of their mortal peril as they comforted thousands of wounded soldiers and dug graves for the dead. Entire hillsides were blanketed with fresh graves caked in red clay; grave diggers were so overwhelmed that unburied bodies were left laying in the open for days until, swollen with gasses, they burst. The humid air stank of rotting corpses and the sickly sweet odor of gangrene.

When General George McClellan's 90,000-man army began its plodding advance up the Virginia Peninsula, Unionists gave free rein to their hopes that Richmond would soon fall, and the war would end with the North victorious. But "Little Mac" never reached Richmond. During the Seven Days battles, waged outside the city during the last week of June 1862, the outnumbered Rebels turned back the Union invaders from the Confederate capital at great cost. Richmond's fifty hospitals did not have space for all of the 16,000 wounded Rebels carried into the city and brought by wagon from the nearby battlefields. At the height of the crisis, storefronts, storerooms, and private homes served as hospital rooms, and Richmond women from all social classes nursed the wounded and maimed—when they weren't anxiously searching the hospitals for their loved ones. Amputees were discharged a few days after surgery to make room for other wounded men; they hobbled around the streets on crutches, mingling with the ceaseless funeral processions.

Manassas had awakened Richmond's citizens to the war's human cost, and subsequent battles had sobered them up from their initial delirium. But the horrors of the Seven Days battles erased the last vestigial romanticism; the capital settled down to the grinding business of fighting a long war. The Confederate surgeon general, S. P. Moore, opened more than twenty hospitals in Richmond, including Chimbo-razo Hospital, the world's largest, with 150 buildings and 100 tents.

True patriots withheld no comfort or luxury that might serve the Confederacy. Richmond women made sheepskin gloves for Confederate officers, sewed burlap sandbags for army batteries, cut up carpets for army blankets and linen for bandages, planted war gardens, donated their spare bedsheets to military hospitals, pawned their jewelry to raise money for the wounded, and volunteered as caregivers. Families took dinner from their tables and gave the food to Rebel soldiers on the march, and then went without dinner. Churches donated their cushions to hospitals for use as beds, and they sold or gave their surplus bells to the Confederate Ordnance Bureau to be melted down and recast into cannons—a 12-ton bell transmogrified into a six-piece field battery.

✶ ✶ ✶

RICHMOND, APRIL 2, 1863

Predominantly female and working-class, nearly a thousand protesters surged through the streets, intent on obtaining food for their hungry families by any means. Food scarcity, magnified by an unusually cold, snowy winter and runaway inflation, had driven them to desperation.

The day before, three hundred of them had met at Belvidere Baptist Church on Oregon Hill and had chosen a delegation to petition Governor John Letcher. But Letcher had rejected all of their requests.

Streaming into the streets with axes and hatchets and crying, "Bread!" the women stormed the Confederate government's provision stores, filling commandeered carts and drays with meal, bacon, and flour. Some of the women already bore the early hallmarks of starvation. "This is all that's left of me," an emaciated eighteen-year-old woman told diarist Sara A. Pryor with a harsh laugh when she caught Pryor gaping at one of her skeletal bare arms.

As merchants watched in horror, the mob rampaged through Richmond's retail district, smashing windows and grabbing shoes and clothing. Watching the scene from the windows of their war prisons, Union captives cheered on the looters.

Governor Letcher issued a stream of orders and threats, and Mayor Joseph Mayo read the mob the riot act. They were ignored; the rioters had gone too far to stop.

The mad pillaging of ten city blocks lasted nearly two hours, until a company of armed soldiers from the Confederate armory led by a Captain Gay marched up Main Street. And then Jefferson Davis emerged from his office and climbed atop an abandoned wagon in the middle of the street. The troops encircled Davis, facing the crowd.

In his short speech Davis sternly advised the women to go home, to stoically bear their privations—which he promised to alleviate if it meant sharing his "last loaf"—and to unite against "the Northern invaders, who are the authors of all our suffering."

Conspicuously removing his watch from his pocket, Davis announced that the women had five minutes to disperse. Then, Captain Gay loudly instructed his men to load their weapons and to shoot to kill when he gave the order.

As Davis coolly studied his watch, the clatter of the ramrods in the soldiers' musket barrels echoed in the now silent street.

Before the five minutes elapsed, the crowd had melted away.

✶ ✶ ✶

Having suppressed the riot, the Confederate government acted to squelch all evidence that it had occurred, fearing disorders would erupt in other cities if the uprising in Richmond became widely known. War Secretary James Seddon prohibited the transmission of any message or telegram alluding to the disturbance. Assistant Adjutant General John Withers requested that the Richmond newspapers avoid making any mention of "the affair," lest there be "misconstruction and misrepresentation abroad."

Most of the newspapers obediently upheld the pretense that nothing unusual had happened in the middle of the Confederate capital. The

Richmond press was practiced at deception; during two years of war, it had evolved into an energetic, sometimes fanatical, propaganda organ of the Confederate government, skilled in inciting public hatred toward the Yankees. The few editors whose scruples—or well-known prejudices against the Davis administration—prevented them from wholly observing Withers's news blackout slanted their accounts. They characterized the rioters as outcasts, army deserters, and "plugs" craving luxury goods, not bread. The *Daily Richmond Examiner* even suggested that enemy agents had incited the mob, a composite of "prostitutes, professional thieves, Irish and Yankee hags, gallows-birds from all lands but our own."

The fabricated reports had their desired effect on most readers. Sally Putnam described the "disgraceful mob" as being made up of "Dutch, Irish, and free negroes" and noted that they had targeted jewelry shops rather than establishments selling groceries. Judith McGuire granted that while some rioters were needy, most were "of the worst class of women, and [there were] a great many who were not in want at all," but who were egged on by "Union influences." Some people, mostly Union sympathizers, remembered the riots differently, but they were decidedly in the minority.

The Confederate government moved so swiftly to suppress reports of the riot because it was painfully aware that it had a food problem, and it feared that Northern leaders might somehow exploit the Rebel vulnerability if they became aware of it. The Richmond Bread Riot, as it came to be called, was in fact the fifth disturbance in the South in three weeks, the others having occurred in Salisbury, North Carolina; Atlanta, Georgia; Mobile, Alabama; and Petersburg, Virginia.

City, state, and Confederate officials took steps to prevent a recurrence. More Confederate troops were sent to Richmond. The city council enacted an ordinance, the Relief of Poor Persons not in the Poor House, which established free-food depots. Over the next year, Richmond would spend more than $150,000 on this and other relief efforts. The Virginia state government hired its own blockade runners to transport Southern cotton to Europe, where they exchanged it for food and supplies. While these actions pre-empted further bread riots, they did not resolve the root problem of the lack of food for the Confederacy's citizens, soldiers—and its growing number of war prisoners. Meanwhile, poultry, salt meat, and even cows and hogs were disappearing nightly, hunger driving otherwise law-abiding citizens to steal to feed their families and themselves.

* * *

The food shortages had struck Richmond harder than any other major Southern city. Besides struggling to absorb up to 100,000 new residents, Richmond also lay within a day's ride of at least 100,000 Confederate Army soldiers, whose needs superseded those of non-combatants. Because of the fighting, fewer acres in Virginia were under cultivation, and most of its food was now imported from Georgia and Alabama; Tennessee and the Carolinas had already been stripped of provisions. The South's ramshackle railroad system, with too few thoroughfares between Virginia and the Deep South, was buckling under the strain of shuttling troops, munitions, food, and supplies, as well as former coastwise shipping curtailed by the tightening Union naval blockade.

Making sure that food reached soldiers, civilians, and war prisoners severely challenged the Confederacy even when shipments arrived as scheduled. When they did not, people went hungry—one month in 1863 just 10 percent of the usual allotment of grain reached Richmond. And all the while, food prices steadily climbed. By April 1863, prices were ten times what they were two years earlier. By late 1863, meat cost twenty-three times more than in early 1861; corn, nineteen times more; flour, twelve times more; sugar, forty-eight times more—and the price of coffee, all of it having to be smuggled through the blockade, was seventy-nine times higher than in early 1861.

To keep a roof over their heads and food on the dinner table, many Richmond women took up "domestic manufactures" like weaving and knitting. They also made pickles and ketchup for restaurants, and wine and jam from berries. The drone of spinning wheels and the banging of looms now heard throughout Richmond were welcomed by Sally Putnam as a return to "the times of our grandmothers." Others left their homes to become government clerks. When her refugee family was unable to live on her husband's salary alone, Judith McGuire went to work in the Confederate Subsistence Department.

Transients of the worst sort were drawn to Richmond by the lure of quick profits and fast living. They were, wrote Putnam, "pernicious characters . . . speculators, gamblers, and bad characters of every grade. . . . Thieving, garroting, and murdering were [their] nightly employments. . . ." Schemes were set afoot every day. Someone bought all the chickens at First Market, and then resold them hours later at a tidy profit. "Strange that men with human hearts can, in these dreadful times, thus grind the poor," McGuire wrote. Richmond's bounty of extortionists, grifters, cheats, prostitutes, quack doctors, counterfeiters, and war profiteers kept the courthouse throbbing with activity. Fortune-tellers did a brisk business, catering to anxious soldiers and civilians hoping to find a glimmer of certainty in the increasingly uncertain future.

Surrounded by violence and death, rival gangs of Richmond boys, most notoriously the upscale Shockoe Hill Cats and the working-class Butchertown Cats, waged furious rock-throwing battles. Sometimes choreographed by invalid soldiers on passes from the hospitals, the rock fights became so pervasive that Jefferson Davis even tried his hand at mediation—without success. A battle of a different order was conducted against the large dog packs that roved Richmond, menacing livestock and people. Citizens retaliated during the so-called Dog War by sending out hunters to kill the dogs. Six hundred fifty canines perished, but the *Daily Richmond Examiner* noted that a large number remained and warned that "a war of extermination" might become necessary.

To many it felt as if the city was sliding into anarchy. The *Charleston Daily Courier* lamented that the old Richmond had been replaced by "a new Richmond, after the moral model of Sodom and New York and Washington." But others attributed the crime and violence to the war and to Richmond's swift evolution into a metropolis. A Richmond reporter hoped that his city might one day become "the Queen City of the South." Still, he missed the "quaint, staid days of 'Auld Lang Syne,'" when every countenance was as familiar as the curbstones, and we felt like one family."

The old churchgoing Richmond of teas and balls, belles and beaus had not vanished altogether, even if it was reduced by poverty and death—and was sometimes invisible amid the mercantile and martial bustle. The young people from the first families of Richmond partied, danced, dined, and attended the theater as though there was not a war being fought less than a day's ride from their doors. If anything, the proximity of death deepened their joie de vivre. But as the fighting intensified, more young officers disappeared from their circle, never to return. Those who did return often bore recent wounds, and sometimes were even missing an arm or leg.

The habitués of this mini-world recognized their fundamental denial of reality. After a ladies' luncheon of gumbo, ducks and olives, oysters, chickens in jelly, salad, chocolate jelly cake, claret soup, and champagne, Mary Chesnut, a South Carolina gentlewoman who was a regular at these gatherings, acknowledged that she and her companions were like "the outsiders at the time of the Flood." She wrote, "We eat, laugh, dance, in lightness of heart!!! . . . [even with] the deep waters closing over us." The *Richmond Daily Whig* also took notice of the phenomenon: "There has never been a gayer winter in Richmond. . . . Go on, good people. It is better to be merry than sad."

This enchanted world lay just blocks from respectable households where gowns, diamonds, pearls, watches, valuable libraries, and silverware were pawned or sold to buy food, and where some families closed off every room but the one in which they lived, slept, and cooked meals over an open fire. And it was utterly alien to the city's waterfront warehouses where war prisoners were slowly starving, and to the nearby

Rebel camps, where thin soldiers scavenged in the countryside for food to augment their meager rations.

Some of those outside the charmed circle resented the stark inequality. Returning home from a military hospital where she had watched a soldier die as his sister "hung over him in agony," Judith McGuire passed a home where there was music and dancing. "The revulsion of feeling was sickening," she wrote. Richmond's revelry reminded her of Paris's gaiety during the French Revolution and of the Brussels ball held the night before Waterloo.

$$\star \quad \star \quad \star$$

RICHMOND, MAY 1863

Saddled, riderless, and with his dead owner's empty boots laid across his saddle, Little Sorrel was led by a groom through Richmond's streets. Behind Thomas "Stonewall" Jackson's horse trailed a mile-long procession that included the Stonewall Brigade, Jefferson Davis and his cabinet, generals, and ranking naval officers. Stunned and heartbroken, thousands of people lined the streets and filled Capitol Square to pay their respects to the Confederate general. Jackson's tactical nimbleness had determined the outcomes of a dozen battles, including Chancellorsville, where he was accidentally shot by his own men.

The pallbearers reverently carried Jackson's casket into the House of Representatives and placed it on a bier before the Speaker's chair. More than 20,000 people filed past, placing lilies of the valley and other spring flowers on the snow-white Confederate flag spread over the casket. "The whole people stood bare-headed and mute," wrote T. C. DeLeon.

The South's grief was a catharsis for all it had borne and lost during two years of war. Enigmatic and idolized, Jackson embodied the hopes of many Southerners, hopes that now began to fade. Even before his death, portents of the Confederacy's mortality could be read in the lengthening roll of Southern defeats and lists of the dead. At Antietam, the Union army had stopped Lee's planned invasion of the North, then New Orleans had fallen, and at Shiloh the Confederacy lost one of its brightest stars, General Albert Sidney Johnston, as well as the battle. "What a noble army of martyrs has already passed away! I tremble for the future," wrote McGuire.

A week before Jackson's death, Union general George Stoneman had led a cavalry raid within five miles of Richmond, throwing the city into a bell-ringing panic. Mary Chesnut, who later became Richmond's best-known wartime diarist, burned her diary, fearing that her candid observations on the Confederacy's crises might somehow serve Union propagandists. "I confess I lost my head," she wrote in her new diary. When the alarm had subsided, it was reported that none of Richmond's batteries had had friction primers for igniting the cannons' main charges; Stoneman could have stormed into the city without a single Rebel cannon being fired.

Abraham Lincoln's Emancipation Proclamation of January 1, 1863, was a far more serious blow. Theoretically, it freed slaves in all states in rebellion against the Union, while also inviting blacks to enlist in the Union Army. Of course, on a practical level it was unenforceable, but its symbolism was powerful: it broadened the war's objective of preserving the Union into a moral crusade against slavery. The proclamation was also a brilliant strategic triumph, dooming the Confederacy's hopes of forming an alliance with England or France; both nations had abolished slavery and would never take sides in a war over slavery against a nation dedicated to emancipation.

Undoubtedly the Confederacy's defeat would mean emancipation, but for the time being the "peculiar institution" remained intact throughout the South, although cracks were appearing behind the façade of normalcy. Richmond's 12,000 slaves and 2,500 free blacks experienced no change in their condition. An 1863 Richmond slave auction brought $3,800 apiece for a female cook and a young black girl, and Southern newspapers continued to report public floggings and to advertise large rewards for runaways. Yet, more slaves were running away—even longtime servants who sometimes absconded, too, with their masters' money, silverware, and jewelry. Insubordination and lawlessness increased. Mary Chesnut and her husband sent all of their servants back to their home in South Carolina and hired new servants in Richmond after a series of troubling incidents, culminating in a slave serving them breakfast while drunk. While Lincoln's manifesto did not improve the lives of Southern blacks, it inspired them to hope that their lives might soon improve.

* * *

By autumn 1863, Richmond's buildings and people were showing the effects of two years of war. No mechanics remained to repair the broken gates and locks, leaking roofs, rotted wood, and crumbling railings. John B. Jones, the War Department clerk, grumbled in his journal that he and other middle-class citizens were "in a half starving condition. I have lost 20 pounds, and my wife and children are emaciated to some extent." Others could not afford the inflated prices for clothing—a suit now costing $700, and boots, $200—and made do with what they had. "We are a shabby-looking people now, gaunt and many in rags," wrote Jones, adding that there would be food and clothing enough for everyone "if we had a Roman Dictator to order an equitable distribution."

While Richmond's citizens might have been unhappy with their government over the scarcity of food and goods and the receding prospect of victory, they wholeheartedly hated the Yankees. "The most dreaded, the most hated of all beings was the 'Yankee,'" wrote Sally Putnam. "Little boys . . . treasured revenge for those who oppressed their fathers. Little girls learned to dread and fear the Yankee above all tame or wild animals." A Chimborazo Hospital matron related how a woman boasted that she kept a pile of "Yankee bones lying around her pump so that the first glance on opening her eyes would rest upon them."

The Yankees had invaded their homeland, killed their young men, and brought hunger and want to their doorsteps. The attitudes of ordinary people and soldiers defending their Southern homeland hardened into a settled malice toward all Yankees, living and dead, captive and free. A war begun almost lightheartedly, with the expectation of it lasting a few short months, had become a prolonged slaughter. For the South, the possible outcomes spanned a narrow range from minimal survival to utter ruin.

2

The POW Archipelago

"The arrangement I have made works largely in our favor. We get rid of a set of miserable wretches, and receive some of the best material I ever saw."

–Confederate colonel Robert Ould,
upon exchanging hundreds of political
prisoners for captured Rebel soldiers

FRANCIS LIEBER'S INTEREST in warfare took root as he watched Napoleon Bonaparte's conquering legions march through the streets of his native Berlin in 1806. His fury over the French invasion of his homeland inspired him to later join the Prussian army, where he served under Prince Gebhard Leberecht von Blücher during his epic attack at Waterloo. Lieber was just fifteen years old.

Barred from Prussian universities for liberal activities, Lieber attended the University of Jena in Saxe-Weimar, obtaining his doctoral degree in 1820. Seven years later, he immigrated to the United States, where he became a founder and editor of the *Encyclopedia Americana*. Lieber taught history and political economy at South Carolina College, where his treatises on civil liberty and political institutions laid the foundations of academic political science in the United States. In 1857 the German scholar was appointed to a chair at Columbia College.

Noting the confusing, often contradictory pronouncements made by Union commanders when the war began in 1861, Lieber proposed a code covering matters such as the treatment of prisoners of war, spies, hostages, and partisans—subjects that had always been subordinate to the primary objective of winning battles. Lieber recognized that a standard of conduct would relieve individual commanders of the burden of making rules for these aspects of their operations, and free them to bring the war to a quick conclusion. This was Lieber's dearest wish, as he had three sons in military service, one of them wearing Confederate gray (and later killed in battle at Williamsburg, Virginia).

An earlier Lieber treatise on guerrilla warfare had introduced the thinking of the clean-shaven, paunchy sexagenarian to General Henry "Old Brains" Halleck, the Union Army's general-in-chief. When Halleck read Lieber's arguments for a code of conduct, he appointed a committee to write one and named Lieber to serve on it. He was the committee's only civilian member, and he wound up writing the code nearly single-handedly.

The Union Army issued the *Code for the Government of Armies in the Field* as General Order No. 100. Better known as the Lieber Code, it was the first known codification of rules for land warfare and became the basis of future conventions, including the Geneva Conventions. Lieber called it "our little pamphlet . . . short but pregnant and weighty like some stumpy Dutchwoman when in the family way with coming twins."

Its 157 articles were comprehensive; its subjects included martial law, prisoners of war, hostages, partisans, spies, and prisoner exchanges. Regarding prisoners, the code prohibited their punishment solely for their being "a public enemy, nor is any revenge [to be] wreaked upon him by the intentional infliction of any suffering, or disgrace, by cruel imprisonment, want of food, by mutilation, death, or any other barbarity." Yet the code permitted calculated reprisals against war prisoners for just cause: "All prisoners of war are liable to the infliction of retaliatory measures."

The code accomplished Lieber's goal of establishing a rational, humane policy for the treatment of war prisoners, thereby freeing commanders to concentrate on tactics and strategy. But the Lieber Code was remiss in one important respect: it did not propose a system for exchanging captives.

✷ ✷ ✷

At the beginning of the war, little attention was given to holding and exchanging prisoners. Because both combatants anticipated a short, decisive conflict, improvised accommodations and exchanges were deemed sufficient. After a battle, the combatant generals often informally agreed to exchange prisoners on the spot. When they were left with hundreds of captives on their hands, Union commanders sometimes telegraphed nearby states, seeking temporary accommodations for them.

But by early 1862, some Union officials were contemplating a formal prisoner exchange system. Already, the Union and Confederacy were having difficulty paroling and exchanging the growing number of captives before their care and feeding became a burden.

President Lincoln, however, refused to negotiate with the Confederacy on prisoner exchanges, believing this to be tantamount to recognition of the Confederate States of America as a sovereign entity.

But with pressure mounting from war prisoners and their families for a formal exchange cartel, Lincoln grew concerned that enlistments would decline if potential recruits feared they could be imprisoned indefinitely if captured. In February 1862, the president sent General John Wool, the doughty Mexican War veteran, to meet with Confederate general Howell Cobb, the former U.S. House Speaker and Treasury secretary.

Wool and Cobb turned to the only extant American cartel model, the one used by U.S. and British authorities during the War of 1812. That system was a vestige of the "civilized" warfare of maneuvers and tactics waged in eighteenth-century Europe between small, professional armies. These wars, for the first time in history, were fought predominantly with firearms, not brutish edged weapons, for limited objectives, with relatively low casualties and few disruptions of civilian life. The era of limited wars and small, professional armies ended with the Napoleonic Wars, which were waged by large armies of mainly conscripts and volunteers, but prisoner exchanges were still conducted as before. During the War of 1812, British and American captives were released at designated sites according to a rigid scale of equivalents: a major general was equal to forty privates or seamen; a colonel, to fifteen privates; a captain, six privates; and a noncommissioned officer, two privates.

The British-American cartel was a vast improvement over the Revolutionary War's lack of systematic exchanges—it having been an insurrection and not a war between sovereign nations. Patriot prisoners were sometimes swiftly paroled and other times held indefinitely (although a well-placed bribe could hasten their release) in makeshift jails and prison hulks—the *Jersey* being the most notorious of the disease-ridden ships moored in British-controlled U.S. harbors. The British in fact murdered, abused, and starved their captives. But the Americans consciously adopted a "policy of humanity" toward their war prisoners, wrote historian David Hackett Fischer, both as a moral statement and to pre-empt a retributive spiral. For two centuries, this was a bedrock U.S. principle during wartime.

At their meeting, Cobb and Wool agreed to adopt the 1812 cartel model. But Union war secretary Edwin Stanton vetoed their agreement, finding fault with the parole provisions. The negotiations were discontinued.

And then the spring and early summer of 1862 ushered in the biggest battles of the war to date. After Shiloh and General George McClellan's campaign on the Virginia peninsula, Confederate prisons were bursting with Northern captives. Their relatives and friends beseeched the U.S. government to obtain their release.

General Robert E. Lee's goodwill gesture of paroling wounded Union prisoners was the catalyst that brought together Confederate general D. H. Hill and Union general John A. Dix to reopen discussions about a prisoner exchange system. In July 1862, they agreed to the same cartel provisions that Stanton had rejected in February. This time, the war secretary did not object.

The cartel stipulated that future captives would be paroled within ten days at designated sites on the James River, and near Vicksburg, Mississippi. Parolees were forbidden to take up arms again until they were officially exchanged, a process that usually took months. In the cartel's early days, Union parolees were sent home. But when many of them failed to return to duty after being formally exchanged, the Army began requiring them to await exchange in special holding camps. From these camps, some parolees were dispatched to fight Indians in Minnesota under General Lew Wallace, who later wrote *Ben Hur*.

The first exchanges occurred along the James River below Richmond on August 3, 1862, when 3,021 Union prisoners were traded for 3,000 Confederates. The cartel was a godsend to the Confederacy, which could now parole the captive Yankees and not worry about feeding them. Weeks before Generals Hill and Dix signed the agreement, food shortages and concerns about security were beginning to develop in the Confederate war prisons. But now prisons were emptied in Macon and Atlanta, Georgia; Mobile and Tuscaloosa, Alabama; Salisbury, North Carolina; Charleston, South Carolina; and Lynchburg, Virginia. Richmond's prisons remained processing centers for Eastern theater captives, but their numbers declined sharply.

In March 1863, Colonel Robert Ould, the Confederate exchange commissioner, watched with satisfaction as Rebel guards escorted hundreds of Union captives to a train headed for the James River exchange site. The parolees were a mixture of enlisted men and political prisoners that Ould was happy to be rid of. "The arrangement I have made works largely in our favor," Ould wrote to General John Winder, commandant of the Richmond prisons. "We get rid of a set of miserable wretches, and receive some of the best material I ever saw."

Ould had entered public life in 1859 amid the blaze of newspaper publicity surrounding the sensational murder of Philip Barton Key. The U.S. attorney for the District of Columbia and the son of Francis Scott Key of "Star-Spangled Banner" fame, Philip Key was shot dead on a street near the White House. No one disputed the fact that Key's killer was New York congressman Daniel Sickles, a Tammany Hall politician and good friend of President James Buchanan. Key and Sickles had been friends until Key had an affair with Sickles's young wife, Teresa, driving Sickles to seek vengeance.

Buchanan appointed Ould to succeed Key, and urged him to go easy on his old friend Sickles. But Ould prosecuted Sickles for murder. However, the jury, persuaded by Sickles's attorney—the future war secretary, Edwin Stanton—found Sickles not guilty by reason of temporary insanity. It was the first acquittal on that ground in the United States. Sickles went on to win military glory and lose a leg as a Union general at Gettysburg. Ould became a Confederate Army judge advocate and, in 1862, the Confederacy's exchange commissioner.

Ould the commissioner conducted himself as uncompromisingly as had Ould the district attorney. He gave no ground when negotiating with Union exchange commissioners, and he was unsympathetic toward the Union captives even when the exchange cartel was running smoothly during late 1862 and early 1863. In a letter to Colonel A. C. Myers, Ould wrote that if trains were needed to carry food to Confederate troops and to war prisoners, "pay no regard to the Yankee prisoners. I would rather they should starve than our own people suffer. I suppose I can safely put it in writing, 'Let them suffer.'"

Indeed the Confederacy's chain of war prisons at times seemed expressly designed to induce suffering. The Rebel prison system, wrote historian William Hesseltine, was "the result of a series of accidents. . . . Prisons came into existence, without definite plans, to meet the exigencies of the moment." During the war, more than 195,000 Union captives were imprisoned in the thirty or more Southern war prisons—many no more than provisional holding facilities that closed when their unsuitability became obvious, or when threatened by Union armies. Few were built for the express purpose of holding war prisoners. Moreover, no single person oversaw the Rebel prison system until the war had nearly ended. As a result, resources were withheld, dispersed, or squandered, and starvation and abuse flourished when they needn't have.

The Confederate prisons were jury-rigged from former tobacco warehouses, factories, fairgrounds, stables, city jails, mills, and, occasionally, a slave pen or a cotton shed. When nothing better could be found, the prisoners camped in open fields, as they did in Blackshear and Thomasville, Georgia. In the South as in the North, the captives were segregated by rank to prevent officers from leading enlisted men in uprisings. The Union officers were held separately at Libby Prison in Richmond, and in the war's later stages, at Macon, Georgia, Charleston, South Carolina, and Danville, Virginia.

Officer and enlisted captives were segregated, too, in the Union's twenty-two prisons: officers on frigid Johnson's Island in Lake Erie, and the rest in prison camps in Ohio, Indiana, Illinois, Delaware, New York, and Maryland. Nearly all of the prison camps in the North were built to hold Rebel prisoners—215,000 total during the war. In October 1861, Colonel William Hoffman of the 8th U.S. Infantry became the system's commissary general. He was so dedicated and efficient that the job became his for the war's duration. Hoffman frequently inspected the prisons to ensure that all of the Rebel prisoners had shelter and adequate food and clothing.

Because the Lincoln administration had rejected the idea of a civilian commission to improve military hospital hygiene, the U.S. Sanitary Commission came into being in 1861 as a volunteer advisory panel to the Medical Bureau. But with the Union Army's phenomenal growth from a standing force of 16,000 to a juggernaut of a million or more, the commission's duties rapidly surpassed the modest responsibilities that were proposed and rejected in 1861. In 1863, after the commission had amassed information about army hygiene and military hospitals and proposed a wholesale reorganization of the medical system, its plan was adopted by Congress. The commission also arranged convalescent care, shipped food to soldiers and war prisoners, and manufactured and distributed blankets through 32,000 auxiliary societies.

It recommended improvements large and small at the Union war prisons and by persistence often obtained them. The commission persuaded Union officials to install a sewer in 1863 at Camp Douglas, a prison near Chicago, that reduced dysentery. But when the commission complained that the 10,000 Rebel prisoners at Point Lookout,

Maryland, were living in drafty tents and proposed new barracks, War Secretary Edwin Stanton expressly forbade their construction. He was in a vindictive mood because of the Union prisoners' poor treatment in the South.

A wide gulf lay between the living conditions of captive Yankees and Rebels. Most Confederate prisoners lived in conditions that their counterparts in Richmond could only dream of. Indeed, Rebel prisoners were fed and clothed better than they were in the Confederate Army, and in many cases, better than they were in their own homes before the war. "I want nothing; I have everything that [my] heart could wish except my freedom," Jonathan Musgrave of Virginia wrote from Camp Chase, Ohio, to friends in the South. John A. Carson heartily agreed: "We have nothing to do but eat and sleep. We have plenty to eat and drink, and a very good bed. We have no reason to complain."

Instead of being systematically robbed when they entered prison— as were Union captives in Richmond—each Rebel prisoner was issued a blanket (two in wintertime), a shirt, a coat, shoes, and trousers. At Camp Morton, Indiana, the prisoners spent much of the day roaming the camp's capacious grounds, and at night they slept inside barracks, two to a bunk. They received the same generous ration distributed to Union troops: 14 ounces of bread or 16 ounces of corn meal; 14 ounces of fresh beef or 10 ounces of pork or bacon; and beans, rice, sugar, coffee, tea, vinegar, potatoes, and molasses. They were permitted to sell the food they did not eat, and to use the proceeds to purchase green vegetables.

Lieutenant Thomas Sturgis, an adjutant at Camp Morton, observed the result of this conscientious care and feeding: "The sight of a sturdy Confederate strolling about with Uncle Sam's U.S. branded between his shoulders was not uncommon." The disparity between Union and Rebel prisoners did not escape the notice of the Union captives. A pale, hungry, and raggedly dressed Captain J. W. Chamberlain watched a group of just-released Confederate parolees march past Libby Prison and sourly remarked that they were "fat and hearty, and well clothed."

Later, Colonel Hoffman reduced the Confederate prisoners' portions because so much was being wasted. And when he learned that some Confederate officers were obtaining good clothing and boots that could serve as Confederate Army attire when they were paroled, he ordered the substitution of inferior-quality clothing and footwear.

Hoffman also revised the policies at the Rebel prison at Camp Chase, Ohio, after learning that the Confederate officers were permitted to stroll the city streets, stay in hotels, and keep their slaves with them. When citizens complained about the extraordinary liberties and indignant Republican officials demanded the slaves' emancipation, Hoffman ended the furloughs and freed the slaves.

But Rebel prisoners also died every day of chronic diarrhea, pneumonia, and disease. Many were ill when they entered prison, weakened by hard marching on lean rations or suffering from unhealed battle wounds. As a result, any contagious disease outbreak was usually deadly. Smallpox at the prison in Rock Island, Illinois, claimed 331 lives in one month. At Camp Douglas, 260 prisoners died in three weeks. Elmira, New York, had the highest mortality rate of the Union prisons, 24 percent, sometimes with up to forty-three prisoners dying in a single day, sparking a brief boom in cemetery plots. Even so, Elmira's death rate was 5 percent to 10 percent below the mortality rates at the worst Confederate prisons.

* * *

In July 1863, the Union government officially suspended all prisoner exchanges, although they had actually stopped two months earlier. During the ten months of the cartel, nearly 200,000 captives from both sides had been exchanged, and by the summer of 1863, Richmond's prisons held no more than four thousand Union soldiers.

The chain of events that led to the cartel's collapse began with President Abraham Lincoln's Emancipation Proclamation six months earlier, which Jefferson Davis denounced the next day as "the most execrable measure recorded in the history of guilty man," and an attempt to "interfere with our social system." The Confederate president blackly prophesied a host of disasters following in the proclamation's train: either the extermination of the slaves, the exile of Southern whites, or permanent separation of the Northern and Southern states. Davis announced draconian measures. "I shall . . . deliver to the several State authorities all commissioned officers of the United States that may hereafter be captured by our forces in any of the States embraced in the proclamation"—meaning anywhere in the Confederacy. The captives would face state criminal charges of inciting slave rebellions, and if convicted, they could be put to death.

While Southern leaders were accusing Lincoln of trying to foment a slave insurrection, the Union Army stepped up the recruitment of blacks eager to fight under the banner of emancipation. (By fall, 50,000 blacks would wear Union blue.) The alarmed Confederate Congress then declared that captured white commanders of black soldiers would be neither exchanged nor returned to Southern state jurisdictions, but would be court-martialed on the field and face summary execution. Nor would captured black troops be exchanged, for the Confederacy could not bear the idea of exchanging a former slave for a Rebel soldier. Black captives would instead be handed to the states in which they were captured and sent into slavery.

The resolution struck a nerve in the North. Unlike in the South, where the war was euphemistically regarded as a struggle for states' rights, most Northerners believed that slavery was the root cause. The Confederate resolution was condemned as a violation of the cartel and the Lieber Code, which stated, "The law of nations knows no distinction of color, and if an enemy of the United States should enslave and sell any captured persons of their army, it would be a case for the severest retaliation, if not redressed upon complaint."

The Lincoln administration cancelled an exchange of officers scheduled for May 25. After further deliberation, on July 13, War Secretary Edwin Stanton formally suspended the Union's participation in the cartel—just as new Yankee captives from Gettysburg and Vicksburg were pouring into Richmond. To continue to exchange prisoners under the South's new proviso, wrote Stanton, would be "a substantial abandonment of the colored troops and their officers . . . a shameful dishonor to the Government bound to protect them."

On July 31, Lincoln issued General Order No. 252: "For every soldier of the United States killed in violation of the laws of war a rebel soldier shall be executed, and for every one enslaved by the enemy or sold into slavery a rebel soldier shall be placed at hard labor on the public works. . . ." Privately, the president harbored strong reservations about killing innocent Rebel captives for their government's actions—it was "a terrible remedy," he conceded to black abolitionist Frederick Douglass. (While Confederate soldiers sometimes executed captured black Union troops on the battlefield, the Union never retaliated in kind.)

In addition to the crisis over the black war prisoners, the fragile cartel had also been jeopardized by the Confederacy's refusal to exchange fifteen hundred men captured while raiding in Alabama and Georgia. Nor would the Confederate command parole or exchange 3,000 men from General Robert Milroy's division who were captured in June 1863, as retribution for Milroy's tyrannical six-month reign in Winchester, Virginia. The Rebels had also imprisoned Union surgeons and chaplains, who were supposedly exempt from captivity.

Moreover, the Confederacy had blatantly violated the wholesale parole of its Vicksburg garrison after its surrender to General Ulysses S. Grant on July 4. Under the terms of their paroles, the 30,000 Confederate captives were free to go home, but they could not fight again until they were formally exchanged for Union captives. But the Rebel command had immediately returned many of the soldiers to active duty. When confronted with this transgression, Colonel Ould had silkily asserted that the Rebels had been exchanged, even when they had not. Vicksburg parolees fought at Chickamauga ten weeks later.

The Confederacy and the Union blamed one other for the cartel's breakdown. "All the obligations imposed on us as to the treatment of prisoners and exchange . . . have been fulfilled on our part with entire and scrupulous good faith," Confederate war secretary James Seddon insisted to Jefferson Davis, "while the course of our enemies has been marked by perfidy and disregard of their engagements and the dictates of humanity." Ethan Allen Hitchcock, the writer-philosopher Union general who oversaw all prisoner exchanges, wrote in the *New York Times* that the Confederacy could no longer be trusted to abide by the cartel's rules. Captured black soldiers must be "treated with that humanity which is due to all other troops in like circumstances according to the laws of civilized warfare," and must not be subjected to "barbarous practices." Colonel Ould professed to be "heartsick" at the termination of the cartel, but he wrote, "I have no self-reproaches. . . . I have struggled in this matter as if it had been a matter of life and death with me."

Now began a collateral cycle that boded ill for Union captives everywhere. As soldiers continued to enter the prisons without any reasonable prospect of parole or exchange, the warehouses and stockade camps became more densely crowded than ever. By late fall of 1863, more than 13,000 Union prisoners were incarcerated in Richmond alone, including 1,200 in Libby Prison and 6,000 to 10,000 enlisted men at the nearby outdoor prison in the James River, Belle Isle. The captives' rations slipped to semistarvation levels, and dysentery and pneumonia spread with frightening speed.

Concern grew throughout the North and South over the mounting human cost of the cartel's breakdown.

3

Inside Libby Prison

"The whole secret of making it endurable consists in having some-thing to do. Something to do, something to do at stated hours, making one forget where he is, is the secret."

–Colonel Frederick Bartleson,
Libby prisoner 1863 to 1864

WHEN REBEL BATTLEFIELD commanders in 1861 and 1862 began sending Yankee captives to Richmond, the Confederate capital was struggling with a population boom and lacked places to put the prisoners. While evaluating their options, Confederate authorities hit upon the idea of commandeering the spacious warehouses and tobacco factories in Richmond's famed tobacco district along the James River waterfront. Among the dozen or more buildings seized by the Confed-eracy was a warehouse that had recently been purchased by fifty-four-year-old Luther Libby.

Three identical, three-story brick buildings joined together as one entity, the warehouse was erected between 1845 and 1852 by John En-ders, a founding member of Richmond's tobacco aristocracy. Adjacent to the Kanawha Canal and James River, it was designed as three ware-houses with common interior walls but no communicating doors, so that Enders could lease the three units separately. Enders did not live to see the warehouse's completion; he fell from a scaffold to his death dur-ing construction.

In 1861, Luther Libby, a Maine native and Richmond businessman, purchased the building and invited his son George into his prosperous ship chandlery and grocery business. But it was an unpropitious time to buy a warehouse on Richmond's waterfront; in March 1862, General John Winder informed Libby that the Confederate government was expropriating his warehouse. Libby was given forty-eight hours to remove his goods. The sign, LIBBY AND SON, SHIP CHANDLERS AND GROCERS, which father and son had raised on the property and which they left behind, became one of the first sights to greet Union captives as they marched to the prison from the Richmond railroad depot. Later, the words LIBBY PRISON were plastered on the wall above the second-floor windows overlooking Cary Street. Through no fault of his own, Luther Libby's name became a byword for misery, abuse, and starvation.

With six chimneys protruding from its broad, sloping roofs, Libby Prison was bounded on the north by Cary Street; by Canal Street on the south; and Twentieth Street to the west. A 50-foot-wide vacant lot bordered the warehouse's east side. Its three cellars, separated by brick walls like the upper rooms, were nearly invisible from Cary Street, but could be entered from Canal Street, where the ground fell steeply toward the nearby canal and river.

Because the warehouse occupied its own city block, apart from other buildings, Confederate authorities believed it would be relatively easy to guard. They made it more secure by whitewashing the dark exterior brick wall up to the second-floor windows so that a would-be escapee descending a rope from a window to the ground would make a vivid target for the armed sentries.

The Rebels piped water from the James into the captives' living quarters, hauled in stoves for cooking and heat, and cut doorways through the walls between the three buildings so the prisoners could mingle (but then quickly nailed them closed to *prevent* mingling). But little else was done to make the warehouse habitable. Like all of Richmond's improvised prisons, Libby was not so much transformed into a jail as emptied of its goods and restocked with captive Yankees.

Libby Prison was the entry point for more than 125,000 Union prisoners sent to Richmond during the war, and tens of thousands of them remained there for months until they were paroled. The prisoners were never permitted to go outdoors for fresh air or exercise, and so they ate, slept, and spent every hour of every day in the warehouse's six upper rooms—each one 105 feet long, 44 feet wide, and 8 feet high—three on the second floor, and three on the third. The three ground-level rooms served other purposes: respectively, a prison hospital, a communal kitchen, and the headquarters offices for Richmond's prisons.

Lieutenant Charles Carroll Gray, an assistant surgeon, described Libby Prison's living areas as "large, low, damp, dirty, cold & dark," and suffocating hot in summertime. Each floor had a water closet—literally a closet with a trough used as a toilet. The river water served every purpose, including bathing, which was done in a large sink under a faucet. From the upper floors' south and west windows, the captives could savor their one reliable pleasure, a splendid view of the James River, its long bridges, and its islands, notably Belle Isle, where the tent roofs of the imprisoned Union enlisted men poked above the railroad trestle that disappeared into the lush, rolling hills stretching toward Petersburg.

Eighteen months after Libby Prison opened, twelve hundred captives lived in densely packed rooms, bumping into one another at every turn. When nighttime came, they lay on the bare wooden floor "spoon fashion," turning over on command when the ache in their bony hips and ribs became unbearable. Lacking blankets, they shivered as wind, rain, and snow blew through the barred but paneless windows. As their numbers grew, their rations shrank to tiny portions of corn bread, rancid bacon, and bug-ridden bean soup, occasionally supplemented by boxes of food from the North and purchases made at the city market by the Rebel guards with the prisoners' hoarded cash.

But for all their discomfort and gnawing hunger, a worse enemy was despair. They fought it with all the energy, cunning, and imagination that they possessed in this great crisis of their lives. While captivity crushed some weaker spirits, most of the officers displayed a fierce resilience to the hardship that threatened their bodies, minds, and souls. They labored to maintain a vestige of civilized life and to keep their minds active. In the process, they created a miniature world.

*   *   *

Colonel Frederick Bartleson described Libby as "a large hotel, where there were bulletins issued from time to time, advertisements of articles lost." Prison life wasn't intrinsically exciting—it was the opposite—but in Libby, wrote Bartleson, there was "an excitement about this kind of living which was peculiar . . . always rushing to and fro." This was by design; the captives battled boredom as though it was death itself. Boredom could initiate a gradual decline that proceeded by stages from apathy to depression to that final destination. Thus, either consciously or unconsciously, prisoners followed the prescription revealed by Bartleson to his wife, Kate: "The whole secret of making it endurable consists in having something to do. Something to do, something to do at stated hours, making one forget where he is, is the secret."

As in any prison, in peace or wartime, Libby's daily routine was minimal, challenging the captives to fill the long hours as best they could. At 6 AM, reveille was announced by a "drum band" of black Union prisoners. It was followed by a quick wash at the faucet in the cold, yellow-tinged water. The lucky dozen or so who owned toothbrushes brushed their teeth (and wore their prize possession in a buttonhole of their shirts). Then came "Old Ben's" stentorian announcements of the morning's headlines—or his unique interpretation of them, anyway. On the street outside the prison, the black newspaper vendor peddled four publications, which were printed on half sheets of brown, dingy paper. The captives bought two or three of them from the guards for a quarter apiece, and the copies were sold and resold, read and reread, until the smudged print was indecipherable. While the newspapers were receiving their first readings, "The General," one of the black prisoners from Libby's west cellar, fumigated the rooms with burning tar, chanting, "Here is your nice smoke without money and without price."

Another important early-morning ritual, besides breakfast and the first of the two daily head counts, was the pursuit and extermination of lice. Captain William Wilkins thought it "a comical sight to see a room full of naked men . . . intently examining with the keenest interest, his trousers or shirt." The "graybacks," as they called the vermin—their nickname, too, not coincidentally, for Confederate soldiers—"literally rained through the cracks of the building upon us," one prisoner reported. Despite the captives' zealous countermeasures, the vermin maintained the upper hand. Some prisoners, either indifferent, lazy, or too sick to care, simply let the lice rule.

Except for a second meal and the occasional visiting Southern dignitary—to whom prominent Union prisoners were pointed out "as if they were in a zoological garden," the captive *New York Times* reporter Junius H. Browne irritably noted—the day's remaining hours crawled by until lights out at 9 PM.

* * *

And so the prisoners organized every conceivable activity to occupy themselves. Willard Glazier described a typical scene in a room at Libby: officers huddled around a chessboard marked out on the floor, moving carved soup bones while their fellows laid bets of money, clothing, even food, on who would win (a Captain Wilson reputedly could beat anyone in Libby with his back to the board, directing a third party to move his pieces). Another group energetically rattled dominoes made from discarded bones or sent by relatives, while others played checkers with brass buttons. And everywhere, it seemed, the captives played poker and seven-up, wagering a spoonful of salt or a pinch of pepper. In the so-called bonton room, the domicile of the highest-ranking officers, barbers trimmed beards and shaved faces.

In a "moot court," complete with judge, jury, and lawyers, phantom acreages, horses, dogs, and cows changed hands; mock bankruptcies were declared; and imaginary marriages failed. "Nearly every man of the number has failed in business, and a large percent have been divorced," Glazier wrote of the proceedings. Occasionally the court put an officer on trial for being an annoyance, as was a Captain Leed, found guilty of bathing after 10 PM.

Those with artistic proclivities wrote, composed, drew, and carved. Carving was easily Libby's leading solitary pastime. With penknives, the prisoners hewed soup bones into spoons, pipes, chess pieces, rings, crosses, napkin rings, and "all manner of articles, useful and ornamental." They carved initials and unit names into walls and column posts. A Captain Fisher sculpted a bust of Abraham Lincoln in a wall. From boxes sent to them from the North, the captives made chairs, tables, stools, and blanket racks. They fashioned clothes hooks from barrel hoops, and lamps out of empty cans, using pork fat for tallow and wisps of cotton for wicks.

Some busied themselves with covering the walls with graffiti—THE UNION MUST AND SHALL BE PRESERVED, UNITED WE STAND, DIVIDED WE FALL, and other patriotic slogans—which the Rebels periodically whitewashed, only to see the slogans reappear. Others made pen and pencil sketches on every level surface, and composed songs, prayers, and poetry. One such composition, "The Prisoners' Song," opened with heartbreaking optimism: "Come brother prisoners join me in song, / Our stay in the prison will not be long." The irony-tinged "Soldier's Prayer" began: "Our father who art in Washington, / Abraham Lincoln is thy name; / Thy will be done in the South as it is done in the North. . . ."

Classes were held day and night. "Anyone unfamiliar with the American character would have thought the Libby verged on becoming a university," wrote Captain Bernhard Domschke of Wisconsin. Many of the officers in civilian life had been doctors, teachers, editors, merchants, civic leaders, and lawyers—of the latter, there were forty, it was said. "The hitherto idle prisoners are *students* now," Chaplain Charles

McCabe wrote to his wife, Kate. The most popular classes were mesmerism (as hypnotism was then known), military tactics, history, and phonography, or "Pitman shorthand." But the prisoners also studied French, Spanish, Latin, Greek, rhetoric, English grammar, arithmetic, algebra, religions, geometry, and science. At McCabe's mess, "we made it a rule that none should have anything to eat at all until he could ask for it in French"—which the twenty messmates were all studying, with McCabe receiving peripatetic tutoring from Chaplain Louis N. Beaudry of the 5th New York Cavalry, who was also the editor of the prison's weekly newspaper, the *Libby Chronicle*. They learned Shakespeare by heart, and deepened their knowledge of bird-watching, a perilous undertaking because one could be shot by guards for standing near a window.

In an era when speeches and debates were regarded as high entertainment, the Libby debating society, the Lyceum (sometimes called "Lyce-I-see-'em," a pun on the ubiquity of lice), was held in the highest esteem. Its members convened in a circle on the floor, "like Indian Chiefs at a war council," according to Captain Robert T. Cornwell, and debated topics important and trivial. The discussions, wrote Colonel Federico Cavada, were conducted "with intense enthusiasm, sometimes even with political virulence, and not seldom with very bad grammar." After each debate, the society determined which side had won. Through this process it was decided that intemperance was a greater evil than war, and the death penalty should be abolished—unusual findings from alcohol-deprived soldiers who had seen so much death. There is no record of which side won the Lyceum's debate of whether black males over twenty-one should be permitted to vote.

* * *

Before there were classes, lectures, and entertainments, there was the Richmond Prison Association, established in 1861 by Union officers and a New York congressman, Alfred Ely. Ely had ridden out from Washington to watch the war's first battle, at Manassas, Virginia, only to be captured. In Libby, he became president of the association, whose purpose was "mutual improvement and amusement"; its defiant motto was "Bite and be damned," another allusion to lice. A year later came the Libby Prisoners Club, whose bylaws emphasized proper order and decorum, equitable distribution of provisions, and punishment for those who took more than their fair share.

By 1863, a prisoner hierarchy no longer existed because there were simply too many captives. For the greater part, distinctions in rank were ignored; the harsh living conditions and the men's separation from their regular units made them more concerned with their personal welfare than with observing military formalities. Only the loosest military organization persisted: a ranking officer was nominally in charge of each of the six rooms, and captives from the Union Army's Eastern Department were separated from the Western Department prisoners, with a colonel at the head of each.

A notable exception to the absence of a prisonwide hierarchy was the "indignation meeting," at which common grievances were aired before a committee composed of three representatives from each room. A recurring subject was the thefts that were rampant in Libby. Tin containers, forks, spoons, and small food stores vanished with such aggravating regularity that one mess inscribed on its coffee pot the words: "To borrow is but human; to return's divine." The representatives also passed judgment on fellow officers. A Captain Forbes, who distributed goods sent by the Union government, was censured for failing to issue a consignment to the captives. Forbes had auctioned the goods, after first supplying his friends.

✴   ✴   ✴

At ten o'clock on Friday mornings, the prisoners sat on the floor in a circle around Chaplain Louis Beaudry, speaking in low voices as Beaudry arranged the slips of paper in his hands. When the chaplain began reading aloud, the officers fell silent, for this was one of the most highly anticipated events of the week: the "publication" of the latest edition of the *Libby Chronicle*. At once entertaining, newsy, witty—and fearfully perishable, too—the *Chronicle* guaranteed that its weekly subscription price was within everyone's budget: "One moment's good attention, invariably in advance. These terms complied with, the news will be forwarded postage free." Lacking a printing press, Beaudry possessed the lone copy, written by many hands on the backs of letters, scraps of paper, and bits of cardboard. It was a medley of announcements of events, programs, classes, lectures, and debates; mock-heroic poems; character-building exhortations; anecdotal accounts; riddles; tongue-in-cheek aphorisms; complaints, trivial and real; actual prison news; and actual war news. The tone was deliberately brisk and light—a reflection of the captives' determination to remain upbeat and engaged.

Among the announcements made during the *Chronicle*'s seven-issue life span during September and October 1863 was one for an upcoming "Bone Fair," where carvers were invited to exhibit their "worked bones" and vie for prizes to be awarded by three judges. There was an involved account of a recent mock trial, and a notice for a new phonography class, which subsequently received two hundred sign-ups. Colonel Federico Cavada offered to teach Spanish and lecture on his native Cuba, while Colonel Louis Di Cesnola, a Sardinian educated in Europe's best military schools, gave lessons in army tactics. And there was a long account on Major John Henry's lecture on "mesmerism," as hypnotism was then known. Henry's classes were deemed by the *Chronicle* to be one of the "three great lights" of Libby, the others being the *Chronicle* itself, it immodestly noted, and the Lyceum. "This is indeed the Augustan Age of Libby Prison," proclaimed the *Chronicle*.

It had words of praise for Chaplain Charles McCabe, known as the "singing chaplain" because he burst into song in his fine, deep voice whenever his comrades were gloomy. McCabe had bought dozens of books and circulated them among the prisoners, transforming Libby into "one of the best literary institutions of Dixie," the *Chronicle* declared with cheerful hyperbole. The books' literary quality ranged from dime novels to *David Copperfield*, *Bleak House*, and *Loiterings of Arthur O'Leary*.

The newspaper was also a forum for advice and complaints, poetry and jokes. Captives were counseled to hide their greenbacks from
the Rebels "between the soles of shoes, in folds of garments, between
the toes," and in their hollow regulation buttons. (To hide a bill in a
button, prisoners pried up the button rim, peeled back the cap, tore
the greenback in two, laid the pieces in the button, and replaced the
cap.) The close quarters, overcrowding, and deprivation understand-
ably bred quarrels and feuds, prompting the *Chronicle* to gently suggest that prisoners try to divest themselves of meanness, selfishness,
and "snappishness." In the same issue, an irate contributor railed
against secondhand smoke: "At least four hundred stinking pipes pol-
lute the air most villainously." One of the weekly "conundrums"
posed the question, "Why is our soup in Libby like the stuff of which
dreams are made of?" The answer: "Because it is a body without substance." And then there was a poem by a homesick prisoner, "To My
Wife": "I think of thee when morning light / Comes struggling e'en
to me, / When waking thoughts mar visions bright, / I think of thee,
I think of thee."

The *Chronicle* was discontinued when Beaudry and the other chaplains were paroled. Beaudry carried out the *Chronicle* files, along with
123 concealed letters written by his comrades. The *Chronicle* is the most
accurate surviving snapshot of life in Libby Prison during the early fall
of 1863. Unlike the captives' journals, which were subjective and private, the *Chronicle* was truth-tested each week by hundreds of men
whose "subscriptions" were bought with their attention.

* * *

Vigorous men in their twenties and early thirties for the most part, the
officers enjoyed playing practical jokes. They roughhoused like boys
to burn off pent-up energy, their chronic exhaustion and hunger
notwithstanding. Whooping gangs of "raiders" periodically rampaged
through the rooms, upsetting chairs, scattering chess pieces, and dis-
rupting classes, chased by volleys of curses.

During the head count one morning, some of the high-spirited young officers played an elaborate prank on Libby Prison's diminutive twenty-one-year-old clerk, Erasmus Ross. When Ross finished counting in one room, twenty to thirty captives wriggled through a flap they had secretly made in one of the former connecting doorways that the Confederates had subsequently sealed—in time to be counted again in the next room. All day long, Ross counted and recounted the captives, never coming up with the same total twice. At one point, he ordered the guards to herd all twelve hundred prisoners into the lower east room, and then counted them as they exited the room in single file. Finally, as the afternoon shadows began to lengthen, Ross decided to average the contradictory head counts that he had taken, and to leave it at that. Ross announced that before the war he had worked in a large New York counting house and, since the beginning of the war, in the prison system—without ever a discrepancy until this day. "Gentlemen, it is due you that I acknowledge myself beat out, for I cannot unravel this enigma." To which the prisoners roared, "Bully for the Yankees!"

Before the Rebels announced lights out at 9 PM, the prisoners sought release and distraction in games, dancing, and other entertainments. One night, officers pretending to be zoo animals paraded through the rooms, heralded by torchbearers and a band playing "Yankee Doodle" and other popular songs on tin plates, cups, kettles, combs, and split quills. Acting the part of an "elephant" were four officers with a blanket thrown over them and two protruding sticks representing tusks; the "camel" was a man sitting astride another man's neck; and the "bear" was a prisoner on his hands and knees. On other nights, the captives danced reels to fiddle music in the kitchen. There were "cockfights" between officers who had been "bucked"—their hands bound and pinioned at the elbows by a stick slid behind their knees. Another favorite game required an officer to stand bent over in the middle of the room, encircled by other prisoners, with one of them shielding his eyes with a hat. Someone would dart from the circle and slap the victim's posterior, prompting him to whirl around quickly and try to identify his assailant—who, if correctly singled out, would then take his turn inside the circle. One officer entertained himself and his comrades by catching mice and teaching them to run on a treadmill made from a tin can. There were theatrical productions of Shakespeare's plays; Broadway-style minstrel shows performed by prisoners with burnt-cork-smudged faces, cracking "myriad dirty jokes . . . mugging and dancing"; and elaborate vocal concerts featuring soloists, duos, trios, and a "grand chorus," singing French, Hungarian, German, Scottish, and Irish songs.

After lights out, the captives often carried on for an hour before settling down to their fitful sleep on the floor. On the quieter nights, they would try to stump one another with riddles, such as, "How does Libby differ from another public institution in Philadelphia?" The answer: Philadelphia has "a Northern home for friendless children," while Libby was "a friendless home for Northern children." Other nights, lights out was the signal for lunacy. The captives howled and shrieked, bawled songs at the top of their voices, and pounded the floors. They frenziedly flung shoes, chunks of corn bread—anything at hand—until they exhausted themselves. When morning came, they spent the first hour retrieving their belongings.

* * *

The cry "Letters from home!" invariably precipitated a mad scramble to the upper west room. No event in Libby stirred up as much excitement as the irregular mail calls, held whenever a "flag of truce" boat bearing letters arrived at City Point, on the James River south of Richmond, from Union-occupied Fortress Monroe. A Union officer perched on a rafter beam and shouted out the name of the letter recipients and, when someone replied, he flung a letter in that direction. Anyone receiving mail regularly was envied nearly as much as the "well-off" captive who owned a toothbrush, two silver spoons, and an extra plate. Aching for word from home, the captives usually read their letters while rooted to the spot where they received them, sometimes rapturously, or, if the news was bad, while weeping. Captain Alonzo

Keeler of the 22nd Michigan remained downcast for days after reading about his wife's serious illness and slow recovery. "This is a poor place for mental trouble," Keeler conceded in his journal.

Total candor was impossible because the Confederates censored all mail, and permitted the prisoners to discuss only domestic matters. Each letter was limited to one page. This was placed in an envelope addressed to the letter recipient, which went inside a second envelope addressed to the commissioner for exchange in Richmond. It was there that a censor read the letter, and if he found nothing objectionable, it was forwarded in the inner envelope to Fortress Monroe for delivery to the addressee in the North. Incoming mail went through the same censorship apparatus. Any letter-writer so foolish as to criticize the Confederates could expect to be banished to the dungeon.

The captives skirted the censors by smuggling out letters with newly paroled prisoners, or by writing messages in "invisible" ink—citrus juice, usually—on seemingly innocuous letters. Recipients could read the notes written in citrus juice by applying heat to the paper. After the exchange cartel's suspension, when only chaplains, surgeons, and the sick and wounded were sent home, the captives increasingly resorted to the invisible ink messages.

Lieutenant George Grant made a special flourish beneath his signature whenever one of his letters to his sister Mary contained a hidden message. In a note scrawled in lemon juice on the back of an ostensibly lighthearted letter describing a session of Libby's mock court, Grant instructed Mary to hide cash in the next box of food and clothing that she sent him: "When you forward the box, put the $10 in the coffee. It will be all right." She was to acknowledge having read his secret note by "drawing a line under your signature" in her next letter. In Lieutenant Grant's papers, his hidden message is clearly visible in brown-tinged letters, explained by the words written in another hand, presumably Mary's: "Held Over the Fire." In her reply to her brother, Mary's signature is underlined.

The Rebel censors discovered the deception when a Union officer, writing in regular ink, thoughtlessly instructed his wife: "Now, my dear, read this over, and then bake it in the oven and read it again." The Rebels, of course, did just that, and discovered the captives' secret. The Confederates thereafter heated outgoing mail before sending it on, and the prisoners reluctantly abandoned their ruse.

* * *

Later in life, some of the more devoutly religious captives would reflect on their confinement with fondness, and not the revulsion that their former comrades harbored until their dying day. One of them was Chaplain McCabe, who regarded his ordeal as "a blessing" because it brought him closer to God. Another was Captain David Caldwell, who wrote, "In some respects, [a] very pleasant eight months of my life." They and others treasured for the rest of their lives the miniature New Testaments that a Northern Bible society sent to the Richmond prisoners. For these captives, the months in Libby Prison tested and reinforced their Christian faith. "In our communing with each other, we could feel it was good even to be in prison, while Jesus reigned with us there," wrote Caldwell.

Caldwell became a temporary chaplain when the Rebels released all of the regular chaplains, leaving the captives' spiritual welfare in the hands of the only Richmond minister who deigned to visit them regularly, Catholic bishop John McGill. Left to their own devices, Caldwell and other spiritual-minded captives tried to fill the void. The Rebel guards mocked the prisoners' improvised services and the presumption that God would listen to their sacrilegious prayers. It was to discourage such gatherings that the prison authorities had nailed shut the doors they had made to connect the rooms.

* * *

Of all the calamities short of death that could befall a captive, none was worse than the punishment of being sent to the "dungeon," located in the gloomy recesses of Libby's middle cellar. Two classes of prisoners suffered banishment there: hostages for Confederate prisoners facing execution in the North, and those being punished for infractions ranging from spitting on the floor to attempting to escape.

The cells were so dark that the prisoners could not read, and sometimes so crowded that they were forced to stand all night long. Often put in irons, they shared a reeking open privy in the middle of the room. Their only physical exercise consisted of taking three steps, turning around, and repeating the process. Corn bread and water was their fare, although their comrades upstairs slipped them food through holes made in the kitchen floor. A. C. Roach described the pitiable sight of dungeon castaways "standing on tip-toe, their bony, skeleton-like arms outstretched," while their comrades lowered contraband beans, rancid beef, and crusts of stale bread.

Bad as were the food, darkness, and crowding, there was something even worse in the cellar—swarms of rats. Bold and beyond enumeration, the rats that overran Richmond's waterfront and Libby's cellars inspired the prisoners to call the prison's low regions "Rat Hell." It was common for a new arrival in the dungeon, while groping his way in the dark, to stumble, fall, and experience the horror of having rats scramble over his face and hands. One captive awakened at night to find a large rat sitting on his head. Eventually, desperation drove the prisoners to trap the rats, cook them on their stoves, and eat them. "If I possessed a pen of living fire I could not paint the horrors of the week I spent in that Libby dungeon," wrote St. George Rathborne.

In the west cellar next to the dungeon, twenty black Union prisoners lived in squalid misery. As Libby Prison's slave laborers, they performed every menial task: swabbing floors, toting slops, cutting wood, cooking, and loading and unloading putrefied bodies that were on their way to the cemetery. For the most trivial infraction, a black prisoner might be stripped, tied to a barrel, and lashed thirty to forty times with a horsewhip or cat-o'-nine-tails. At night, their screams and moans sometimes awakened the officer captives in the upstairs rooms. After a particularly long, gruesome whipping heard by all the prisoners one night, they learned that a black captive had been caught trying to escape and had received more than a hundred lashes.

Invisible though they were in Libby Prison—and their very invisibility cloaked their activities—the black captives served the white Union officers in a vital way unknown to the Rebel authorities. Because they routinely came into contact with black servants from the city, they were able to obtain information about Confederate troop movements and the activities of Richmond's Unionist underground. The intelligence, passed to them by pro-Union blacks, was then transmitted to the captive officers. Prisoners contemplating escaping especially prized the information.

4

The Defiant Colonel

"Even criminals guilty of the blackest crimes are not, among civilized people, confined for any length of time on insufficient food."

—Union colonel Abel Streight, in his letter of
complaint to the Confederate war secretary

RICHMOND, MAY 16, 1863

This wasn't how Colonel Abel Streight and the fifteen hundred men of his Independent Provisional Brigade had wanted to enter the enemy capital. The hulking, blue-eyed commander and his raiders might have imagined themselves as conquering heroes. Instead, hostile Southerners lined the streets to watch them force-marched by armed Confederate troops through Shockoe Bottom to prison.

Streight had planned to wreak havoc on supply depots and railroads in Alabama and Georgia, and to lure Union sympathizers down from the northern Alabama hills. But in every sense, his raid had been a failure.

Failure was a new experience for the thirty-four-year-old Indiana general. With a common-school education, Streight had left his boyhood New York farm to find success at nineteen as a contractor, and later as a businessman in the frontier city of Indianapolis. When the war began, he was a leading citizen of Indiana's capital, the owner of the Railroad City Publishing Company and the New Lumber Yard. He had done it with a unique combination of brains, aggressiveness, and intimidating physical presence—at 6 foot 2 and 225 pounds, he was half a foot taller and 80 pounds heavier than the average American man. A photo of Streight wearing his Union Army dress blues shows a large man whose dark, bushy beard contrasts oddly with his clean-shaven upper lip and pale eyes.

An abolitionist, Republican, and Lincoln man, Streight interviewed the president immediately after his election in 1860, and on the eve of the war, he wrote a pamphlet advocating the forcible restoration of the South: "A rivulet of blood, spilt at this time, will prevent rivers of it in the future." Streight's hawkish views, combative nature, and political connections won him an appointment in September 1861 as commander of the 51st Indiana Infantry Regiment, with the rank of colonel.

In lacking military experience, Streight was no different than hundreds of other high-ranking officers in the North and South who received volunteer regiment commands at the beginning of the war. He learned quickly while leading his regiment at the siege of Corinth, Mississippi; at Perryville, Kentucky; and at Stones River, Tennessee, in December 1862.

In January 1863, he was sent home because of a medical condition described as "congestion of the cerebellum"—probably migraine headaches. When he recovered, Streight was uninterested in rejoining his regiment in camp. He craved action.

Streight wanted to emulate the brilliant Confederate cavalry general, Nathan Bedford Forrest, known for his slashing raids behind Union lines. In March 1863, Streight proposed to his friend and superior, General James A. Garfield, a bold plan for a mounted raid into "the interior of the South"—although he had no experience as a cavalry commander. While operating along the Mississippi-Alabama border the previous July, Streight and the 51st Indiana had made contact with Alabama Unionists and enlisted forty recruits. "Never did people stand in greater need of protection," Streight told the future twentieth president. By conducting such a mission, he "could do more harm and our cause more good" than he had accomplished during the past year. Garfield authorized Streight's raid, giving him command of the 51st Indiana and three other Midwestern infantry regiments, plus two companies of Tennessee cavalry. The Independent Provisional Brigade was born.

Streight's plan was risky, but the decision to mount the raiders on mules, and not horses, was a grave mistake. The Union command believed mules would be more sure-footed in the rugged northern Alabama hill country and better able to withstand the rigors of a two-hundred-mile raid. And, the thinking went, while mules were slower than horses, the raiders had little need for swift mounts, as they were unlikely to meet any opposition.

In the after-action report that he submitted more than a year later, Streight ruefully concluded: "I am convinced that had we been furnished at Nashville with 800 good horses, instead of poor, young mules, we would have been successful." Because the Union command supplied just eight hundred mules for the fifteen hundred men, Streight had to canvass the countryside to find several hundred more. When they were rounded up, his infantrymen then faced the daunting task of saddle-breaking them.

At last, the troops sailed up the Tennessee River to Eastport, Mississippi, and debarked. As soon as they reached dry land, four hundred of the mules bolted into the countryside. After catching some of the fugitives and finding replacements for some of the others, the sardonically nicknamed "Lightning Mule Brigade," with three hundred men on foot, crossed into Alabama.

Unfortunately for Streight's raiders, the Rebel cavalryman that Streight so admired, General Nathan Bedford Forrest, happened to be in the vicinity. It wasn't long before Forrest was in hot pursuit of Streight's brigade. Forrest's tough veteran cavalrymen relentlessly pressed the Union troops across northern Alabama, guided by the loud braying of the Yankees' mules. The Yankees and Rebels fought a series of small, brutal battles. Although bloodied by Union ambushes, Forrest's men invariably regrouped and resumed their pursuit, not permitting their quarry to rest. Believing that their wounded comrades were slowing them down, Streight's harried men abandoned them beside the trail with food and blankets. The fitful running fight ground down the slow, balky mules, and they died by the score. Historian Robert Willett Jr. described it as "one of the extreme cases recorded of animal abuse."

The endurance contest ended near Gaylesville, Georgia, where the raiders destroyed the Round Mountain Iron Works—the mission's only tangible achievement. Forrest's cavalrymen pounced on the exhausted Yankees when they stopped for breakfast after marching all night. Some were sleeping so soundly that they could not be rousted to fight, and others dozed off on the skirmish line. Rather than attack, Forrest's men repeatedly circled Streight's raiders to create the impression of overwhelming numbers. The ruse worked. Convinced that he was outnumbered and that his men were in no condition to fight, Streight surrendered. Just twenty footsore mules survived.

After Streight handed Forrest his sword and his men stacked their arms, Streight learned the depth of Forrest's deception. With just five hundred men, Forrest had captured Streight's brigade, which was three times larger. Streight demanded that his arms be returned so that he could fight it out, a proposition that amused Forrest, but that he rejected. In his after-battle report, Streight altered the facts: "We were confronted by fully three times our number."

Swept up with Streight's brigade were escaped slaves who had followed the doomed raiders in the hope of gaining their freedom; Forrest turned them over to Alabama officials. He granted the Union captives lenient surrender terms that included the promise of a quick parole and exchange. But when the captives reached Richmond, Confederate officials renounced Forrest's generous terms.

As they marched through Richmond's streets to Libby Prison, Streight and his Lightning Mule Brigade were taunted and threatened, jeered, and heckled by onlookers. All of the captured brigade members officially entered the Rebel prison system at Libby, but only Streight and his ninety-five officers remained there; the enlisted men were transferred to other warehouse prisons in Richmond and to Belle Isle. The officers supplied their names, ranks, and unit numbers when they entered Libby Prison. Then, in what was known as the "scaling process," the Rebels methodically stripped them of everything of value.

✶ ✶ ✶

With the blessing of Libby's commandant, Major Thomas Turner, a Virginian who left West Point to enlist when the war began, the Rebels confiscated the prisoners' blankets, haversacks, canteens, belts, serviceable clothing, and money. They snatched away the captives' pencils, combs, pocketknives, watches, and anything else they coveted. Meanwhile, an officer menaced the Yankees with a pistol, warning that anyone concealing valuables, or refusing an order, "would be shot down like a dog." The Rebels sometimes subjected the men to minute examinations, even studying their uniform buttons for evidence that they had been pried open and greenbacks cached there. Captain David Caldwell disgustedly pronounced the shakedown to be "a studied, systematic theory of wholesale robbery." John Urban watched a Confederate soldier snatch a prisoner's ambrotype of his wife and, contemptuous of the captive's pleas, stamp it to pieces "with a malignity that would disgrace a South Sea heathen."

It was useless to complain—especially to the hated warden, Richard Turner, a former plantation overseer and no relation to Major Thomas Turner. When a Lieutenant Ballard told Richard Turner that he had "no right" to remove a small piece of "bomb-shell" from Ballard's pocket, Turner roared, "No right!" With a curse, he added, "I'll show you!" and backhanded him. Turner also struck Lieutenant Frank Moran in the face—which was still bandaged and swollen from a battle wound—when Moran raised objections to the thefts. Moran more than got even down through the years by writing countless articles about Libby Prison, each one depicting Turner as a great villain.

Libby Prison's first commandant was Richard Turner's antithesis. Major J. T. W. Hairston got the assignment when typhoid made him unfit for field command. The officer prisoners never again had it so good as they did under Hairston: he treated them like gentlemen. But when he recovered his health, Hairston returned to active duty, and two mean-spirited commandants succeeded him—Swiss-born Captain Henry Wirz, the future notorious commander of Andersonville, and Lieutenant David Todd, a brother-in-law of President Lincoln who energetically baited and abused the prisoners to compensate for his connection to the enemy commander-in-chief.

Richard Turner continued the brutal practices of Wirz and Todd. When Union prisoners hissed the Confederate flag as the Rebels prepared to raise it over Libby, Turner ordered each of them "bucked and gagged," the gag being a steel bayonet placed between the jaws and tied to the head. On Turner's watch, prisoners were routinely kicked, punched, and sent to the dungeon for trivial offenses. A Captain King of the 12th Pennsylvania Cavalry spent two days in the cellar jail subsisting on bread and water because, in Turner's presence, King missed a spittoon with a stream of tobacco juice. Once, four hundred Union captives were forced to stand at attention all night while Confederate troops aimed loaded muskets at them, with orders to fire instantly and slaughter all of them if one of them so much as moved. They survived, never knowing why they were subjected to the ordeal.

* * *

A Libby captive threw down a note to Captain Bernhard Domschke and his men as they were being marched into the prison: "Hide your greenbacks." The Confederates especially prized U.S. folding money, for their own currency was steadily losing value against the Union greenback—worth ten Confederate dollars in May 1863, and equal to twenty by December.

Prisoners concealed their cash in their caps, coat linings, and boots. Others adopted even more creative ploys. George Darby cut a small strip of leather from his boot, wrapped it around a $2 greenback and a $20 Confederate note, and placed it inside his mouth. Another officer filled his pipe with greenbacks, put tobacco on top of it and puffed on the pipe while being searched; no one thought to search the smokeless pipe. Others were not so quick or clever. "Sir, that is not my money," Chaplain Charles McCabe protested when a Confederate took $80 in greenbacks entrusted to him by soldiers to send home to their wives and mothers. "I know," replied the Confederate, "it is mine now."

Some captives meekly handed over their cash and received printed receipts that stated the prisoner's name and the amount taken and included the sunny promise that the sum would be returned upon the captive's release. But one wonders whether Lieutenant Joseph Foley, from whom $32 was taken, or Jacob Thompson, who handed over $416, ever got their money back.

It was well worth the trouble to smuggle in U.S. currency, because the prisoners could exchange it for Confederate cash on the black market operated by the guards and use it to buy extra food. The Confederate dollar had plummeted to shocking depths because the Rebel government manufactured ever larger quantities of scrip to pay its mounting debts, and because it permitted states and cities to issue their own currencies. As a result, a hodgepodge of paper money circulated throughout the South, and the black market for Union greenbacks thrived.

It was a criminal offense to exchange Rebel "graybacks" for Yankee greenbacks, but even the supposedly disapproving Confederate government dealt in the enemy's currency when it could get it and gave it to the blockade runners to buy European goods. Richmond brokers boldly displayed greenbacks in their windows, under the government's nose. "The papers said the rage after our money was a mania," observed Colonel Frederick Bartleson.

In his journal, Captain J. W. Chamberlain described one of his encounters with a guard who dealt in black-market greenbacks. Standing on the stairway to his room, Chamberlain waited for the pacing guard, with his bayoneted musket at shoulder rest, to come into sight. When he did, Chamberlain gave the prearranged signal. The guard stopped, and with the bayonet traced a number on the wall before resuming his march. The number denoted the exchange rate, which was evidently agreeable to Chamberlain because, a short time later, he dropped a wad of greenbacks at the guard's feet as he passed by; scarcely pausing, the Confederate in one swift motion scooped them up and pocketed them. When the guard returned to duty a day or two later, he tossed Chamberlain a roll of Confederate bills.

A few inventive, steady-handed Libby Prison captives recognized opportunity in the guards' illicit currency exchange—an opportunity for counterfeiting. With steel pens, they manufactured $5 bills and traded them for Confederate cash. At nearby Pemberton Prison, Robert Sneden transformed $1 bills to $10 denominations and sometimes to $100 notes; his tools were India ink, a brush of camel's hair, and blood—to get the color right. Careful to always pass the bills to the guards at dusk, when the light was poor, Sneden was never caught.

* * *

After the Rebels had robbed Streight and his officers of every valuable that they could find, they sent them upstairs to their quarters. There the new prisoners were met by cries of "Fresh fish!" A crowd of men peppered them with questions about the war and life on the outside. It was the Libby prisoners' way of welcoming newcomers. They asked, "What's your corps?" "How are we doing in the West?" "Got any papers?" When the new prisoners tried to explain how they came to be captured, the "old rats" jibed, "Oh, yes! We know how it happened. Overwhelmed by superior numbers."

Finally, the new arrivals, dazed by all that had happened to them, were released to stake out their places in the room. Each man was entitled to a 6-foot-by-2-foot area on the floor that he learned to recognize as his by the men on either side of him. Soon enough, the bleakness of their surroundings dispelled any residual shock over their capture and robbery, and they had to fight off a nauseating wave of hopelessness. "It is a sort of unnatural tomb, whose pale, wan inhabitants gaze vacantly through the barred windows," wrote a morose Colonel Federico Cavada after his arrival a few weeks later.

Confederate officials immediately announced that Streight and his men would neither be exchanged, nor classified as prisoners of war. Instead, they would be sent to Alabama to face a host of charges that included inciting slaves to rebel, a capital offense. Although the charges were false—during its running fight across northern Alabama, the Lightning Mule Brigade had had little opportunity to eat or sleep, much less incite a slave insurrection—the Richmond newspapers nonetheless characterized Streight as a great danger to the Confederacy. "He was hated as heartily as if he had been altogether successful," observed Junius Browne, the imprisoned *New York Tribune* correspondent.

But inside Libby Prison, Streight's comrades looked to him for leadership because of his forceful personality and outsized physique. Within weeks, Streight had become president of the new Libby Prison Association, the successor to Congressman Ely's Richmond Prison Association and to the Libby Prisoners Club.

* * *

LIBBY PRISON: JULY 4, 1863

The Yankee officers crowded into "Streight's Room" to celebrate Independence Day. If anything, captivity had made the officers more devoted to freedom, deepened their patriotism, and instilled a quiet pride in their special sacrifice for the Union. While there was little for them to celebrate—the great events at Gettysburg and Vicksburg on this day were as yet unknown to them—it was an occasion for singing the patriotic anthems that so irritated their captors.

More important, this was the day that they planned to unfurl an American flag that had been secretly made by Captain Edward Driscoll of the 3rd Ohio, a former tailor. The Rebels prohibited any display of the Stars and Stripes. So Driscoll had discreetly stitched a respectable-looking flag from donated strips of red flannel, scraps of white cotton, and a blue patch provided by a navy ensign. While the flag was more white and blue than red, one prisoner trenchantly observed, "We had only to think of the battlefields from which we had come to fancy there was more of red than either."

At one o'clock, the prisoners filed into the room, and a Captain Reed climbed into the rafters and hung the homemade flag. The ceremony began with a prayer and a reading of the Declaration of Independence, as the flag switched in the warm breeze that wafted through the paneless window openings, moderating the rising midday heat. Then Colonel Streight, as president of the Libby Prison Association, was invited to speak. During his rousing extemporaneous address, the prisoners cheered, applauded, and stamped their feet.

The noise attracted the attention of a Confederate sergeant, who came upstairs to investigate. At first "bewildered and dumbfounded" by the sight of an American flag on display in the Rebel prison in defiant violation of the prison rules, the sergeant collected his wits and headed straight for Major Thomas Turner's office to report what he had seen. He returned to Streight's room with an order from the commandant to take down the flag. The officers refused, and the sergeant returned to Turner for further instructions. A short time later, he reappeared. Red-faced with embarrassment, he climbed into the rafters, seized the flag, and forbade them to continue their Fourth of July observance.

As the sergeant watched from the head of the stairs, the prisoners, led by Streight and other Libby Prison Association officers, launched into an intensive, hours-long debate over whether to flout Turner's order and resume the ceremony. The debate itself became a de facto observance of sorts, noted Captain David Caldwell. "Really we were having our celebration, and about every speech that had been prepared for the occasion was delivered, if not in full, yet in synopsis." And so when they decided not to continue the ceremony, their point had already been made, and they turned to music, which had not been forbidden. They sang "Rally 'Round the Flag, Boys," "We're Coming, Father Abraham, 600,000 More," and other favorites that they knew would rankle their captors, such as "The Star-Spangled Banner" and "We'll Hang Jeff Davis from a Sour Apple Tree." They concluded with the thrilling patriotic anthem, the one that drove the Rebels to fury—"Yankee Doodle."

And then, two days later, the momentous news of the Union Army's triumph at Gettysburg reached Libby Prison, whipping the captives into a state of tremendous excitement. Gettysburg was a crushing defeat for the Confederacy, for it meant that the Rebel invasion of the North had been stopped cold, and that Robert E. Lee's army was turning back to Virginia. Chaplain Charles McCabe, the "singing chaplain," decided that it was the right occasion to lead the captives in a little-known song whose lyrics he had read in the *Atlantic Monthly* and memorized: "The Battle Hymn of the Republic."

Julia Ward Howe had written the lyrics in 1861 after meeting with President Lincoln in Washington and attending a troop review at a nearby army camp, where the band played the popular army song, "John Brown's Body." Howe's pastor, the Reverend James Freeman Clarke, urged her to write new lyrics for that melody. The new words, she later wrote, came to her the next morning as she lay awake in bed before dawn. She had the presence of mind to write them down before drifting back to sleep.

Inspired by the Gettysburg victory, McCabe taught the song to his fellow prisoners. Soon, Libby's rooms were ringing with "Mine eyes have seen the glory. . . ." It became the Libby prisoners' favorite song, and the one that the Confederates detested above all others. When their captors banned it, the Yankees sang it with more brio than before.

After he was exchanged in the fall of 1863, McCabe sang "Battle Hymn of the Republic" before a large audience in the House of Representatives that included President Lincoln and Vice President Hannibal Hamlin. Moved by McCabe's performance, the listeners rose and joined in the chorus. A delighted Lincoln exhorted McCabe to sing it again. After the war, McCabe went on the lecture circuit, where he told and retold the story of singing "Battle Hymn" in Libby Prison. Julia Ward Howe later said McCabe and the Libby prisoners helped make it the war's signature Union song.

* * *

In April 1863, two Rebel captains, William Francis Corbin and Thomas Jefferson McGraw, returned to their native Kentucky from duty in Virginia to help raise a Confederate regiment. They were captured by Union troops and charged with being spies.

Kentucky was divided in its loyalties, and Kentuckians served in both combatants' armies. Rebel leaders had been so certain that Kentucky would join them that they placed a star representing the state on the Confederate flag. But the Lincoln administration succeeded in preventing Kentucky's secession.

About the time of Corbin's and McGraw's capture, Union general Ambrose Burnside issued Order No. 38, which stated the circumstances under which enemy captives would not be classified as war prisoners, but as "traitors and spies" subject to trial and execution. "Secret recruiting officers within our lines" was one of them.

At a military trial in Cincinnati on April 21, Corbin and McGraw were sentenced to die before a firing squad on Johnson's Island, the Union prison camp for Rebel officers in Lake Erie. Corbin's family desperately tried to commute the sentence. Corbin was a respected church elder in Campbell County, Kentucky, and his relatives were able to collect the signatures of Union sympathizer friends on a petition avouching for the men's character. Corbin's sister, Melissa Corbin, presented the petition to General Burnside and appealed for clemency. But Burnside refused to commute the sentences, agreeing only to forward the petition to President Lincoln without any recommendation. Melissa Corbin traveled to Washington to seek an audience with the president, but Lincoln would not see her.

On May 15, Corbin and McGraw were marched onto the beach at Johnson's Island and blindfolded. The firing squad's volley bowled them into the coffins that had been positioned behind them.

On July 6, as Libby Prison's captives sang "Battle Hymn" to celebrate the Gettysburg victory, a macabre ceremony was conducted in another part of the prison, the first phase of the Rebels' plan to avenge Corbin's and McGraw's deaths. All of the Union captains in Libby were assembled in a hollow square around a table. Before them was a ballot box containing seventy-five slips of paper—each bearing a captain's name. The names of two captains were to be randomly drawn from the box, and these unlucky ones would die in expiation for the executions of McGraw and Corbin.

The captains chose Chaplain Joseph T. Brown of the 6th Maryland, whose name was not in the box, for the wretched task of selecting the victims. With trepidation, Chaplain Brown approached the ballot box. All eyes were on him. Pulses raced, and sweat beaded on foreheads. "He made a short prayer, after which he closed his eyes, and with tears streaming down his cheeks, he drew out two slips," wrote Captain J. W. Chamberlain, one of the seventy-five. A Rebel officer read two names: captains Henry Washington Sawyer of the 1st New Jersey Cavalry, and John P. Flynn of Colonel Streight's 51st Indiana.

As the other officers sighed in relief, guards escorted Sawyer and Flynn to the Libby dungeon, where they were put on a bread-and-water ration pending their executions in a few days. Major Thomas Turner then granted them a two-week reprieve so that they could correspond with their relatives before they died. In a letter to his wife, Sawyer wrote, "I have only the consolation that it is not through anything I have done, or anything I could evade doing, to bring this severe affliction on my family." The letter, reprinted a week later in the *National Intelligencer*, evoked strong public sympathy in the North and caused an outcry for the Lincoln administration to act.

Mrs. Sawyer traveled to Washington, where she was granted an audience with Lincoln and War Secretary Edwin Stanton and made an appeal for her husband's life. The Lincoln administration, in fact, was already working on a plan to stop the executions. They intended to raise the stakes. A few days after Mrs. Sawyer's meeting with the president, the Union government informed the Confederacy that if Sawyer and Flynn were put to death, the captive sons of General Robert E. Lee and General John Winder—General W. H. F. "Rooney" Lee and Captain William Sidney Winder—would also die. Lee and Winder were transferred to Fortress Monroe in preparation for their hanging.

The Confederate government balked. Unwilling to sacrifice a Lee and a Winder for two captains, it suspended the executions of Flynn and Sawyer. Soon their sentences were commuted, and they were released from the dungeon to rejoin their comrades in Libby's upstairs rooms. The Lincoln administration's brinkmanship had won the day. Months later, Flynn and Sawyer were exchanged for Lee and Winder.

The adversaries had been testing one another's willingness to put to death certain classes of prisoners since November 1861, when fourteen captured crewmen from the Confederate privateer *Savannah* were charged with piracy in a U.S. court; a guilty verdict would have condemned them to die. The Confederate Congress promptly removed fourteen ranking Union officers from war prisons in Richmond and Charleston and held them hostage in local jails, ready to kill them if the privateer crew was executed. Reacting to intensive public pressure, the Lincoln administration reclassified the seamen as war prisoners instead of pirates, and the crisis ended.

* * *

No sooner had Libby Prison's captives finished celebrating the Union victory at Gettysburg than they learned that the more than 30,000 besieged Confederate defenders at Vicksburg had surrendered on July 4 to General Ulysses S. Grant. Now, not only was Robert E. Lee in full retreat in Maryland, the Union controlled the Mississippi River; Texas, Louisiana, and Arkansas were severed from the rest of the Confederacy. The prisoners spontaneously launched into a spirited rendition of the "Star-Spangled Banner" and then sang patriotic songs until midnight, pointedly selecting those that Junius Browne described as "obnoxious to the Rebels, and therefore particularly agreeable to us."

In Richmond, the conflation of demoralizing news was "astounding; it was paralyzing. The spirits of the people were deeply bowed down by defeat," wrote Sally Putnam. Mary Chesnut mourned friends who had died in the war. "Nearly all that pleasant company are dead— and our world, the only world we cared for, literally kicked to pieces."

Southern mortification turned into rage when it was revealed that General John Pemberton had surrendered Vicksburg on the Fourth of July on the presumption of getting better terms from Grant on Independence Day. (Grant did generously parole the Rebel captives, rather than imprison them.)

Jefferson Davis exhorted the people to fight harder and not give up: "Fellow-citizens, no alternative is left you but victory or subjugation, slavery, and the utter ruin of yourselves, your families, and your country." One consequence of Vicksburg's capitulation that was felt in Richmond was the virtual disappearance of sugar, grown in now inaccessible Louisiana and Texas; desserts made with sugar became rare treats.

Days after the twin defeats, large crowds of men and women scornfully described in Richmond newspapers as "Irish, Dutch, and Jews" appeared at the office of General John Winder, Richmond's provost marshal, seeking passes to the North. Observing the chaotic scene from his clerk's desk in the War Department, John B. Jones mordantly concluded that the would-be emigrants planned to leave their families behind in Richmond in order to "save the property they have accumulated under the protection of the government."

* * *

Amid the military reversals and rising defeatism, Confederate officials dropped their plans to send the Lightning Mule Brigade's officers to Alabama to face criminal charges. The Confederate exchange commissioner, Colonel Robert Ould, announced that he was willing to parole the Lightning Mule Brigade whenever the suspended exchange cartel was revived. Richmond prison officials were eager to get rid of the troublemaking Colonel Abel Streight.

Streight basked in the enemy's hatred as though it were the highest praise, and "hated them back with an intensity that must have left some margin in his favor," wrote Junius Browne. During his two months at Libby, Streight had relentlessly goaded his captors, fiercely defended his countrymen against bullying Rebels, and emerged as a leader of the prison's vocally defiant faction. When Confederate guards attempted to intimidate him or a fellow prisoner within his earshot, a fellow captive reported, Streight would shoot back, "You dare not carry out that threat! You know our government will never permit it, but will promptly retaliate upon your own officers!" He had been in Libby Prison just one week when he wrote a letter to Major Thomas Turner protesting his being robbed by Rebels in Atlanta, where he lost nearly $3,000 in U.S. funds and $932 in Confederate scrip.

The Confederates despised Streight for having led one of the war's largest Union raids into their homeland, and for his complaints, protests, and the irritating Fourth of July celebration. But he then took hostilities to a new level by complaining about the Libby Prison captives' treatment in a letter to Confederate war secretary James Seddon, and sending copies to the Union and Confederate exchange commissioners, General S. A. Meredith and Colonel Ould. Streight's scorching letter instantly elevated Libby Prison's living conditions to great importance in both the North and the South.

Streight made such an impact because of his high rank and celebrity as a Union raider, his fearlessness in telling the truth, and his journalist's flair for invective. He wrote that the prisoners faced systematic "semi-starvation. . . . Even criminals guilty of the blackest crimes are not, among civilized people, confined for any length of time on insufficient food." Indeed, the rations had shrunk to bare-sustenance levels as more captives jammed into the warehouse prisons, without hope of parole. Each day, the captives received one-quarter pound of "poor fresh beef," a half pound of bread, and a half gill of rice or beans, Streight wrote. Scurvy had already appeared, killing one prisoner. It was not more widespread only because the officers pooled their money—up to $1,000 a day—to buy vegetables at the Richmond market. But the captives' funds, diminished by the Confederates' robberies, were now nearly gone.

In the September 11 *Libby Chronicle*, the editor, Chaplain Louis Beaudry, reprised Streight's letter in full, adding, "It is not likely that the Rebel Secretary of War will condescend to answer Col. Streight's letter. The cry of famishing prisoners cannot enter such delicate ears."

✳   ✳   ✳

In fact, the letter spurred the Confederate government into action. To the captives' amazement, their rations improved, if only briefly. The Rebel authorities went to the trouble of disputing Streight's charges by producing a rebuttal signed by three Union officers, one of them the gray-bearded supervisor of food distribution at Libby, Lieutenant Colonel James Sanderson.

A former New York hotel keeper, Sanderson was the Army of the Potomac's commissary officer and author of the Army cookbook *Camp Fires and Camp Cooking*, a guide to food preparation in the field. "Remember that beans, badly boiled, kill more than bullets" was one of its admonitions.

Captured during the Union Army's pursuit of Lee's retreating troops after Gettysburg, Sanderson now held a position of great responsibility and authority in Libby Prison. He and his assistants had organized the prisoners into twenty-man mess squads, taught them how to cook, and divided the daily rations among them. Disagreeable though the job was, Sanderson "performed this duty with zeal and fidelity," wrote Colonel Federico Cavada. Feeding twelve hundred food-obsessed men was no picnic. No matter what he did, Sanderson was disliked and criticized by one group or another. Some messes accused him of playing favorites when he distributed rations and settled quarrels in the kitchen, while others believed him to be scrupulously fair. Sanderson's superior living standard was a subject of rumor and suspicion. In fact, he owed his privileged status to his Northern friends' generosity and his culinary skills, and not to his having skimmed provisions, as some captives alleged.

Sanderson may have actually believed, as the rebuttal letter stated, that Streight's allegations were "false and unjust" and that the captives received courteous treatment, or he might have been acting on the belief that Streight's letter had impugned his performance as culinary director. But he certainly did not anticipate the rancor that his statement stirred up among his fellow captives. At an "indignation meeting" of prisoner representatives from Libby's six rooms, Sanderson was censured for "gross misstatement of facts," and Streight was praised for attempting to improve living conditions. Thereafter, Sanderson was routinely accused by many of his fellow officers of "flunkyism" and disloyalty to the Union, while his small circle of loyal friends praised his competence, fairness, and generosity, and dismissed Streight as a blowhard.

5

Misery and Retaliation

"In the prison cell I sit, thinking, mother dear, of you
And the dear and happy home so far away,
And the tears they fill my eyes, spite of all that I can do,
And I try to cheer my comrades and be gay."

—A Union captive in Richmond

"This may seem a hard species of warfare, but it brings the sad re-
alities of war home to those who have been directly or indirectly in-
strumental in involving us in its attendant calamities."

—General William Sherman on the subject of total war

IN OCTOBER 1863, the prisoners' rations were cut in half—to a half loaf of corn bread per man, a few ounces of meat, and some bean soup—all of it virtually inedible. The corn bread was hard as railroad iron, and the captives could smell the black-and-blue meat, during its rare appearances, the instant that it arrived in a room. "We can scarcely go near it, much less eat it," complained one hungry officer. The bacon was maggoty, and the soup full of black bugs that had gorged on the insides of the beans and then died, floating to the surface. The *Libby Chronicle* advised the prisoners to get their food "down at once" without smelling, looking at, or tasting it, "just as a shark swallows a dolphin."

The prisoners demanded that they be permitted to do their own cooking, a privilege that was granted. Although censured for having challenged Colonel Streight's allegations of abuse, Lieutenant Colonel James Sanderson remained the prisoners' "culinary director." He made certain that Libby's twelve hundred men received a ration every morning and afternoon, and that they knew how to prepare it. He set the hours when each twenty-man mess could cook to prevent a fierce struggle from breaking out twice daily at the smoking stoves.

It made little difference, though, because the ration continued to shrink. By mid-November, the prisoners went without meat for ten days or more at a time. "Nine ounces of . . . corn bread and a cup of water per day are poorer rations than those issued to the vilest criminals in the meanest States Prison in the Union," complained one officer.

By December, the Union officers were getting 2 ounces of beef about every four weeks. Wrote H. M. Davidson, a Chickamauga captive: "The whole arrangement of these jails was admirably adapted to the purpose of slowly poisoning their wretched inmates. . . . There is no torture so intense as this fierce longing for food."

As they gradually starved, the prisoners obsessively discussed food; they dreamed about it; they planned imaginary dinners and pretended to eat them. "The rich viands and sparkling wines, like those of the banquet recorded in the Arabian Nights, existed only in the imagination of the guests," wrote Captain I. N. Johnston of the 6th Kentucky.

They ate their skimpy rations in small bites, savoring each one. "I try to keep each bite in my mouth as long as I can," wrote Benjamin Booth, but he admitted that his self-control often slipped. After losing 25 pounds in two months, Michael Dougherty of the 13th Pennsylvania Cavalry concluded that it would be better "if they would take us out in an open space and place us before a battery of cannon and blow us to pieces."

The prisoners scorned the Confederates' insistence that the captives received the same ration as Rebel troops in the field. "No army could be kept in the physical condition of General Lee's [army] upon a Libby ration," wrote Captain Johnston. This was true, and the Confederate government quietly made the disparities into policy, while insisting publicly that there were no disparities: soldiers in the field were to get more rations than soldiers "at post," and war prisoners were to receive the same as the latter. While the conscript militiamen who normally guarded the prisoners in theory were as ill fed as the captives, in practice they had the advantage of being able to steal from the Yankees' rations.

* * *

The captives pooled whatever cash they had smuggled into prison to buy food from the nearby city market. The warden, Richard Turner, and his handpicked guards shopped for the prisoners, exchanging the U.S. greenbacks at a profit, and then outrageously overcharging the prisoners. Even as it drained the captives' hoarded cash, this system prompted the Confederates to slash the rations further, hoping to stimulate greater prisoner cash outlays and profits for themselves.

Because the system excluded prisoners without cash, those lacking greenbacks were compelled to sell the guards their rings and watches, even their shoes, to raise money for food. When there was finally nothing left to sell or trade, they simply wasted away. They began to show the symptoms of incipient starvation—dizziness when they stood up too quickly, a ringing in their ears, voices reduced to husky whispers. Frowning over the squares marked on the floor, some chess players now got headaches when they played; a few fainted.

* * *

The last refuge from starvation was the crate of food and clothing sent from the North by a relative, a friend, the Union government, or the U.S. Sanitary Commission. Early in the war, the boxes, carried up the James River on the "flag of truce" boats, reached the prisoners unopened and intact. Captives sent their relatives lengthy, detailed requests. Captain Robert Cornwell's letter to his wife Lydia was as much inventory as correspondence: "1 large canvass ham, 1 hank dried beef, 15 lbs. Sugar, 1 gal. Syrup, molasses, 20 lbs. Cheese, 15 or 20 lbs. butter. . . ." Cornwell went on for ten more lines—cooking utensils, writing materials, clothing, and a bushel and a half of potatoes. Amazingly, he received everything he asked for. If Cornwell and his messmates observed the usual protocol, the box was opened with great ceremony, and each item was presented for all to see and pass judgment upon. These rituals were the climax of the week or month.

But when the Confederates learned that the boxes often contained forbidden greenbacks and other contraband, they began searching them before handing them to the captives. With marlin spikes, the guards ripped them open and sifted the contents for cash. Sometimes their frustration with the tedious hide-and-seek game got the upper hand, and they wrecked the boxes' contents. "Cans of fruit are burst open and probed with a stick—the same stick being used in all cans, whether of pickles or salt. Cheese is split open, bottles emptied, and packages all torn loose and contents emptied," complained Adjutant S. H. M. Byers of the 5th Iowa. A government shipment of warm coats vanished, and Rebel soldiers soon were seen walking around Richmond in blue coats. When the prisoners' bitter protests over the invasive searches reached the Union exchange commissioner, General S. A. Meredith, he accused the Confederates of stealing from the boxes to supply Robert E. Lee's army. The Rebels replied that they were being unfairly vilified for taking legitimate countermeasures against the prisoners' evasions and perfidy, and they sought ways to repay the Union insult to Southern honor.

* * *

Of all the thousands of war captives who passed through Richmond's prison warehouses, none was more famous than fifty-nine-year-old General Neal Dow. Slightly built, gray-haired, and unimposing-looking, Dow was a world-renowned temperance crusader, and the "father of the Maine Law," the first statewide prohibition act in the United States. A celebrated speaker, Dow was Portland's leading citizen: an investor, tannery owner, chief engineer of the city fire department, and mayor of the city. Although reared as a Quaker, Dow turned out to be too combative for Quaker pacifism—as a boy on the Portland, Maine, waterfront, he beat up a monkey that attacked him, and in 1861, at the age of fifty-seven, Dow volunteered for military service so that his ailing son could stay home and mind the family business interests. He was commissioned a colonel and raised an infantry regiment, the 13th Maine. In June 1863, Dow was wounded and captured at Port Hudson, Mississippi. He became an instant celebrity in Libby Prison; fellow prisoners encouraged him to make temperance speeches—even though most of the captives were necessarily teetotalers—and Richmond newspaper reporters sometimes even covered his orations.

For all of his fame, high rank, and advanced years, Dow never became a leader in Libby Prison, although the other captives implicitly trusted the small man in the red skullcap and sought him out for advice. He was one of them: true-blue loyal to the Union, anti-slavery, and harboring "a perfect contempt" for traitors. Revered as he was, Dow could be a potent troublemaker when he believed that an injustice had been committed.

* * *

In late October 1863, Dow learned of the dire conditions on Belle Isle, where enlisted prisoners were held. In letters folded up and concealed in the shirt buttons of paroled surgeons and chaplains, he alerted the Union command to the desperate need for clothing and blankets on the small island in the James River. Soon, a U.S. government shipment of these articles arrived in Richmond, and Confederate authorities, unaware that Dow's smuggled letters had prompted the shipment, chose him to supervise its distribution on Belle Isle.

When he arrived at the prison island on November 6 with the clothes and blankets, Dow was stunned by the misery that he saw everywhere. "All the privations we had encountered at Libby were as joys of Elysium," he wrote. The captives lived in "mildewed, ragged tents," reported Dow, or in "shallow graves" they had dug in the ground.

When it was too cold for them to sleep, they constituted a spectral army: "In the calm cold light of the moon, they could be seen moving about, feebly striking their shoulders with their hands," wrote Lieutenant G. E. Sabre, who had voluntarily transferred from Libby to Belle Isle on the basis of its misleading name. Others slept piled on top of each other in ditches "like hogs in winter," as one prisoner described it, taking turns occupying the outside of the row.

When the James River froze so solidly that it could support wagons carrying heavy loads, a new, terrible morning ritual began at Belle Isle: collecting the bodies of men who had frozen to death during the night, up to fourteen at a time. In November 1863 a Confederate officer on Belle Isle described the island to Chaplain H. C. Trumbull of the 10th Connecticut as "a perfect slaughter pen for your men."

After leaving Belle Isle on November 6, Dow described what he had seen to John Hussey of the U.S. Christian Commission, who had brought the clothing and blankets to Richmond, but had not been allowed on the island. The next day, Hussey shared Dow's impressions in a letter to General Meredith. On November 8, Dow boldly wrote to Colonel Robert Ould, the Confederate exchange commissioner, that the Belle Isle prisoners lived "in a most wretched condition, suffering very much every way," and entreated him to improve their situation. He also requested permission to bring six assistants to help him deliver a second consignment of blankets and clothing to Belle Isle.

Ould responded by prohibiting Dow's return to Belle Isle. He would not be permitted to hand out blankets and clothes from the North when they arrived later in the month. Not only was Dow "utterly inefficient," Ould wrote to General Meredith by way of explanation, "he has shamefully broken his word of honor in not confining himself exclusively to that work in his intercourse with the prisoners." General John Winder instructed Libby's commandant, Major Thomas Turner, to select three "discreet officers" to take Dow's place.

Unbeknownst to Ould, before his banishment from Belle Isle, Dow had also written directly to Union war secretary Edwin Stanton. In his letter, Dow expounded upon the prisoners' hardships in Libby Prison and requested commensurate treatment, down to the "starvation rations," for Confederate captives in the North. "Our captivity is aggravated by the knowledge that rebel officers in our hands have comfortable quarters, abundant and various rations of excellent quality," Dow told Stanton. "The officers here are very earnest that rebel officers, prisoners, may be placed on precisely the same level that we occupy. . . . Such a course may possibly obtain for us some amelioration of our condition."

* * *

The potential repercussions of dwindling food and prison crowding worried General John Winder, and he tightened security in anticipation of uprisings. Even so, he did not believe that his guards, no matter how prepared they were, would be equal to such a crisis. "No force under my command can prove adequate to the control of 13,000 hungry prisoners," Winder warned War Secretary James Seddon.

Fearful that Libby's prisoners might be communicating with outside allies and co-conspirators, the Confederates now forbade the captives to stand within 3 feet of the windows. Guards were ordered to shoot at any prisoner whose face or hands appeared at a window, and they stood with cocked muskets in the street opposite the prison. Scarcely a day passed without shots being fired at prison windows by trigger-happy guards. They "gunned" for the prisoners' heads "pretty much after the fashion of boys after squirrels," wrote Lieutenant Frank Moran. "The whiz of a bullet through the windows became too common an occurrence to occasion remark unless someone was shot." Moran sardonically suggested that more prisoners were not hit because after a few months in Libby, they became so thin that they made poor targets.

Lieutenant George Forsyth of the 100th Ohio was fatally shot in the head while reading a newspaper 10 feet from the nearest window. Fragments of his shattered skull wounded two other captives. A guard shot a prisoner dead as he hung a blanket in the window to dry; Lieutenant Daniel Hammond of the 10th Pennsylvania was shot in the ear as he was washing at a sink. While walking across a room, John Hickey was killed by a city militiaman aiming at a prisoner in the room below Hickey's. The militiaman, who had carried his musket fully cocked all morning, was arrested—a rarity. Hickey's friends vowed to someday "burn him alive at a slow fire." When the prisoners complained to Major Thomas Turner about the guards' conduct, Turner laconically replied, "The boys need practice."

★ ★ ★

General Meredith requested a meeting with Colonel Ould to discuss Dow's letter to Stanton. Rebel prisoners were comfortably housed and well-fed, wrote Meredith, and if the Confederacy could not do the same for the Union prisoners, it should release them. Ould derided the suggestion, as it was "not accompanied with any propositions to release our prisoners now in your hands." He called the allegations of neglect and cruelty "infamously false" and ventured that the "personal honor" of the Union officers who had made them had failed them. "There is nothing in the action of the Confederate Government which gives any sort of countenance to the charge of cruelty or inhumanity to your pris-oners," Ould concluded.

On November 26, Meredith and Ould met at City Point just as ninety-five newly paroled Union surgeons were arriving, prefatory to their being exchanged by the Confederacy for captive Rebel surgeons. After speaking with the surgeons, armed with fresh details about the inhumane treatment that they had witnessed, Meredith sat down with Ould. When Meredith had finished his account, Ould professed aston-ishment and denied everything. The surgeons on the adjoining boat would willingly testify otherwise, Meredith responded, offering to put all of them under oath. Meredith then summoned a Dr. Bowes and questioned him in Ould's presence, followed by naval surgeon W. W. Myers. Both men corroborated Meredith's allegations of abuse. Professing shock and outrage, Ould vowed to question his subordinate officers to identify and punish whoever was responsible.

While Meredith was at City Point, two of the surgeons told him that the Confederate Army also was expropriating U.S. government food and clothing sent to the Richmond prisoners. Having previously read of this practice in smuggled captives' letters, Meredith now ordered the government shipments stopped until he could investigate further.

* * *

In November 1863, several paroled surgeons wrote public statements and affidavits avouching what they had witnessed in the Richmond prisons, and Union newspapers rushed the shocking details into print. The three hospitals where war prisoners received medical care had insufficient medicine, the surgeons said, and on average fifty patients had died each day since Chickamauga—or fifteen hundred a month. "We have over ten percent of the whole number of prisoners held classed as sick men, who need the most assiduous and skillful attention; yet, in the essential matter of rations, they are receiving nothing but corn bread and sweet potatoes . . . unsuitable diet for hospital patients prostrated with diarrhoea [*sic*], dysentery and fever. . . ." And at least five hundred other prisoners who might have benefited from even this low order of hospital treatment were refused it, the surgeons reported.

Because they were denied hospital care until they were at death's door, the surgeons wrote, on one day eleven of sixteen men brought by ambulance to a hospital died over the next twenty-four hours; on another day, ten of fourteen admitted men died. Meanwhile, Richmond's daily newspapers, observed one former captive, usually ended their "abusive editorials by declaring that even such treatment is better than the invading Yankees deserve."

Worse lay ahead, the surgeons predicted. "We are horrified when we picture the wholesale misery and death that will come with the biting frosts of winter." They concluded by observing, "No prison or penitentiary ever seen by them [the surgeons] in a Northern state equaled, in cheerlessness, unhealthiness, and paucity of rations issued, either of the military prisons [Libby or Belle Isle] of Richmond, Va."

Because they were the first Libby prisoners to be released after the suspension of the exchange cartel, the medical men left Richmond stuffed from head to toe with correspondence. Willard Glazier described how "almost every button on the coats of those going home contained a good-sized letter written on tissue paper. . . . The soles of their shoes and boots were loosened and papers put between them. The crowns of hats and caps were ripped apart, filled with letters, and sewed together again."

Widely circulated in Northern newspapers, these letters comprised a new wave of disturbing details about Rebel prisons that built upon the surgeons' reports. Throughout November and December, leading Northern newspapers, including the *Philadelphia Bulletin*, the *New York Times*, *Harper's Weekly*, and the *National Intelligencer*, published gruesome articles based on the smuggled accounts that included illustrations of skeletal men. "The horrors of the *Jersey* prison-ship [the infamous British prison hulk of the Revolutionary War] have been revived in the treatment of our poor incarcerated soldiers," declared the *National Intelligencer*. "These men are starved deliberately. No other alternative is suggested." The fire kindled by Neal Dow's letters and the surgeons' statements became a conflagration.

* * *

While all of this was happening in November and early December, the Confederacy released 189 sick and wounded men from Belle Isle, followed by 300 more. The North reacted with horror when the filthy, emaciated parolees reached the Union lines. Even veteran Union Army surgeons were stunned by the men's appearance when they arrived at the Annapolis military hospital. Assistant surgeon S. J. Radcliffe described their "dangling, long, attenuated arms and legs, sharp, pinched features, ghastly cadaveric [*sic*] countenances, [and] deep sepulchral eyes." Their blackened skin was caked with "loathsome filth," they were covered with lice and "large foul ulcers and sores," and their battle wounds were largely untreated. One of the men who died in transit had received only a field dressing for the fractured skull he had suffered at Gettysburg, four months previously. Many were delirious with fever, and diarrhea was rampant. Some of them were carried to the hospital while dying from pneumonia; others were succumbing to starvation. Eight were dead on arrival at Annapolis; forty-eight others perished over the next ten days. As the sick, wounded, starving parolees expired in his care at the rate of five per day, Surgeon Radcliffe wondered, "How great must be the mortality, then . . . of those still suffering the pains of imprisonment."

* * *

The first demands for retaliation were made in Northern newspapers and at the U.S. Capitol. It was widely believed that the Confederacy would treat Union captives better if Rebel prisoners were deliberately neglected. This was a fallacy. The Confederate prison system was overwhelmed. Southern agriculture and the Rebels' food distribution and transportation networks could not keep up with the needs of soldiers, civilians, and captives. Virginia's crops had been largely ruined by nearly forty-five straight days of rainfall over the summer. Moreover, as the war swallowed up more Southern farms, food production continued to decline, especially in Kentucky and Tennessee. And the North's once ridiculed "anaconda" strategy of girdling and squeezing the South was in fact producing results. The Northern blockade had reduced food shipments to Southern ports, and the Union now controlled the Mississippi Valley, severing the core Confederacy from its cattle- and sugar-producing regions in Texas and Louisiana.

Yet the North remained largely unaware of the extent of the growing hardships in the South. Thus, two issues dominated debate in the newspapers and Congress: whether to restart the exchange cartel despite the South's refusal to release black war prisoners or their white officers; and whether to make Rebel prisoners atone for the Confederacy's abuse of Union captives.

These were wrenching moral questions for Union officials. Even "Old Brains," General-in-chief Henry Halleck, was conflicted. He granted that the captives' treatment was "more barbarous than that which Christian captives formerly suffered from the pirates of Tripoli, Tunis, and Algiers," and worse than the British hulks of the Revolutionary War—"or the 'Black Hole of Calcutta.'" For these reasons he believed that retaliation "is fully justified by the laws and usages of war." At the same time, he found it "revolting to our sense of humanity to be forced to so cruel an alternative." President Lincoln, too, recoiled from the prospect of retributive measures. Most ranking army and government officials, however, did not.

Many newspapers advocated a general prisoner exchange without conditions. "We have no way but to yield. Give them their demands . . . no matter what, if it will only ransom these heroes from the grips of their tormentors while life yet flickers!" exhorted the *New York Times*. The *Boston Daily Advertiser* argued that the cartel's suspension punished the poor men already starving in Libby and other prisons. Whatever the cost, wrote the *Advertiser*, the Union must redeem the war prisoners.

But these arguments for a general exchange were brushed aside as a consensus formed for retaliation. During a Senate debate, it was noted that at the height of the Quasi-War in 1799 and during the War of 1812, Congress had authorized the president to retaliate against, respectively, French and English prisoners, for cruel, barbarous acts against American captives. Former proponents of humane treatment for Rebel prisoners came forward to repent their previous solicitude. Union general Truman Seymour, who had ministered to the Confederate wounded at Antietam, wrote that he had experienced a change of heart after becoming a Rebel captive. It was wrong, Seymour now believed, that "for our rebel prisoners we construct elegant accommodations and admit luxuries, while our people rot with dirt and scurvy." After a Sanitary Commission inspection of Richmond's war prisons confirmed what everyone had said and written, the Union command ordered General Robert E. Lee's captive son, General W. H. F. "Rooney" Lee, to receive "special treatment similar to that which the rebels extend to Union prisoners."

The advocates of retaliation prevailed, and General Ethan Allen Hitchcock introduced punitive measures in the Northern prisons. Commandants banished the sutlers, whose wares had eased the Confederate captives' lives, and they then issued a blanket prohibition against all prisoner purchases. Even as he approved these actions, the sixty-five-year-old Hitchcock, a meticulous, scholarly man who wrote philosophical treatises in his spare time, expressed concern—shared by General Winder, his Confederate counterpart—that the measures might cause desperate prisoners to rise up in camps "where the means of security are very slender." As a precaution, he recommended transferring Rebel captives from camps with minimal security either to state penitentiaries or to island prisons guarded by fortified batteries.

At the same time War Secretary Edwin Stanton took the momentous step of cutting back the prisoners' rations. "You are directed to take measures for precisely similar treatment [to that in Richmond's prisons] toward all the prisoners held by the United States, in respect to food, clothing, medical treatment, and other necessities," Stanton instructed Hitchcock. The reduced ration remained relatively generous, so much so that it was cut again in mid-1864. Even then, it was larger than the daily subsistence of Confederate combat troops. While the Union adopted retaliation as its official policy, it lacked the ruthlessness to deliberately inflict semistarvation on the Rebel captives.

* * *

Fleeting optimism that an agreement might be reached on a general exchange foundered on the Union's insistence that Rebels captured at Vicksburg be returned to their paroles and inactive service. The Confederacy rejected this condition, insisting that the 30,000 troops had been properly exchanged (they were not) and therefore were eligible for active duty. From the Union's standpoint, it was just as well, for a general exchange would have benefited the Confederacy; 26,000 Rebels were held in the North, and 14,000 Union captives in the South. Moreover, Union leaders were beginning to steel themselves for "total war"—described by General William Sherman as "a hard species of warfare . . . [that] brings the sad realities of war home to those who have been directly or indirectly instrumental" in beginning it.

As it was understood by Sherman and his commander and friend General Ulysses Grant, total war meant that captives and civilians were potential combatants, and that crops, livestock, railroads, shipping, and commerce—any and all pillars of the enemy's war effort—could be destroyed. As the Union command's strategy shifted in late 1863 from capturing territory and winning battles in the hope of forcing the Confederacy to seek peace, to grinding down the South's manpower, matériel, and transportation network to win an unconditional victory, other issues, such as prisoner treatment, began to matter less.

Total war advocates questioned the wisdom of resuming the exchange under *any* circumstances. Why, as General Grant would argue, hand over to the Confederacy healthy, rested combat troops ready to rejoin the ranks? With the Union preparing to wage a more primal mode of warfare with modern efficiency, it now mattered less whether, as the *National Intelligencer* suggested, it might be inhumane to refuse "medicines, food, and clothing to the poor creatures on Bedloe's Island, in order that the gaunt and shadowy procession of the Richmond tyranny may have its parallel in New York."

* * *

To counteract the inflammatory stories of captive abuse circulating in the North, in early 1864 the Confederacy launched a public relations counteroffensive in the Richmond newspapers and conducted its own war prison inspections. Predictably, the findings deplored "the utter falsity" of the accusations leveled against the Confederacy. The prisons, wrote Major Isaac Carrington, were reasonably clean; blankets and clothing had been distributed; all of Belle Isle's prisoners were living in tents and had access to "excellent water"; and the captives received the same rations as Rebel troops, although for a short time they had not received a full ration of meat. Hitchcock dryly noted that the report contradicted every shred of evidence that he had seen, as well as the testimony of dozens of former prisoners, and of the surgeons who had treated wounded and ill parolees.

After publishing the report findings, the *Richmond Dispatch* accused Northern newspapers of spreading lies about the Rebel war prisons in order to "blow up the declining war spirit in the North," and to justify the "new cruelties" that the Union had in store for the Rebel captives. The Confederate government, the *Dispatch* assured its readers, was doing all that it could to feed thousands of captives whose needs were burdening a city that was struggling just to feed its citizens. The Yankees could not expect "to fare sumptuously every day. . . . If we are starving ourselves, how can we keep them from starving?" By suspending the exchange cartel, wrote the *Daily Richmond Examiner*, the North had foisted thousands of hungry men upon Richmond. "We certainly cannot find [*sic*] them in victuals much longer. They have already eaten up all our beef and have begun upon the sheep." If the Union refused to take back the captives, the *Examiner* suggested putting them "where the cold weather and scant fare will thin them out in accordance with the laws of nature." The *Daily Richmond Enquirer* wrote that "death on the field of battle [would be] far better than captivity here this winter, and would accordingly counsel them [the Yankees] also not to be taken alive."

The whitewash and newspaper broadsides notwithstanding, Southern leaders knew the truth about their war prisons. In an internal memorandum, assistant adjutant general E. Pliny Bryan wrote that his inspection of Richmond's prisons had exposed "shocking" conditions, yet he also acknowledged, "In our present condition, I suppose we cannot do much better."

* * *

No Union general was more hated in the South than Benjamin Butler, and in the fall of 1863 he was appointed commander at Fortress Monroe for the second time. During his first command tour there in May 1861, Butler had outraged Southerners by designating captured and fugitive slaves as "contraband of war" and therefore, the Union's to set free. His action helped prepare the way for the Emancipation Proclamation.

A year later, after aiding in New Orleans's capture and becoming its military governor, Butler placed the city under martial law. Loathed for his official conduct and for his troops' blatant thievery, Butler was given the nickname Spoons by New Orleans residents, because of the citywide disappearance of family silverware. Butler did not enhance his popularity when he billeted his officers in the homes of Confederate officers, emptied the banks of Confederate funds, dispersed mobs by deploying field artillery, and closed a newspaper that refused to publish his proclamation to the city. Less known were Butler's acts of beneficence, such as feeding 32,400 impoverished people at a cost of $50,000 a month, despite his army's unreadiness for a humanitarian mission.

Butler was especially reviled, however, for hanging a man who tore down the American flag at the New Orleans Mint, and for his infamous "woman order." Reacting to complaints that New Orleans women were spitting on soldiers, and that one woman had dumped dirty water from a balcony onto Admiral David Farragut, Butler issued General Order No. 28, subjecting any female who insulted a Union soldier to the same treatment as a prostitute. For his insult to Southern womanhood, he became known throughout the South as Beast Butler. On December 24, 1862, Jefferson Davis denounced Butler as an outlaw, and condemned him to be hanged on the spot if captured. Butler's subordinate officers, if they were apprehended, were to share their commander's fate.

A successful Boston criminal lawyer in civilian life and the son of a cavalry commander that had served with Andrew Jackson at New Orleans, Butler had always been interested in the exchange cartel's legal aspects. When he returned to Fortress Monroe in 1863, he asked War Secretary Stanton to permit him to attempt to restart the prisoner exchanges. Stanton balked initially; but when December arrived and no prisoners had yet been exchanged, he grudgingly gave Butler permission to proceed.

On December 17, Butler was named special agent of exchange, and General Meredith became his subordinate. Butler was authorized to conduct man-for-man exchanges on the condition that black and white troops were treated identically. He chose to temporarily ignore the Confederacy's flagrant parole violations.

The Confederacy strenuously objected to "Beast Butler" becoming the Union's official exchange commissioner and, in protest, refused to parley with him. Butler's appointment, wrote the *Daily Richmond Examiner*, was "a filthy action." But when the Union would not revoke Butler's commission, the Rebel government, eager to resume prisoner exchanges, became more conciliatory; new exchanges meant bringing home badly needed Confederate troops, and sending away the Union prisoners who were consuming Richmond's provender. If achieving these objectives required it to accept Butler as the Union commissioner, the Confederacy was willing to swallow its pride and negotiate with him.

Butler's efforts resulted in a man-for-man exchange on Christmas Day that sent home 520 Richmond prisoners. But the Confederacy stubbornly refused to change its position on exchanging black prisoners, and the brief thaw ended just three days later.

* * *

As hopes for the revival of prisoner exchanges waxed and waned, Colonel Ould announced that the Confederacy would no longer deliver boxes containing food and clothing sent from the North to the Richmond captives. It was a hard blow. The prisoners would now have to live exclusively on the rations they received from the Confederates— except for the fortunate ones with cash to buy goods from the local market. Days later, Major Thomas Turner abruptly closed this final avenue by forbidding his staff to buy food for the prisoners.

In his December 11 letter to the Union command announcing the suspension of box deliveries, Ould said that the final straw had been the U.S. Sanitary Commission's new practice of addressing its boxes of blankets, clothing, and flour to "Our Starving Soldiers in Richmond" or "Our Brave Defenders in Libby Prison." This certainly annoyed the Rebels to the point of refusing to deliver the boxes, but it was a flimsy excuse; the issue of prisoners receiving goods from up North and the Rebel guards' increasingly intrusive searches and brazen thefts had been building to a crisis for two months. Ould's action, characterized as "childish" and uncivilized in the Northern newspapers, created an instant crisis inside the Richmond war prisons. "We may now look for stiff times," remarked Lieutenant Cyrus Heffley. The captives quickly consumed every leftover scrap of half-rotten potato or cabbage, and every stale bread crust found on the grimy floors. Hungarian-born Colonel Emeric Szabad wrote that merely staying alive now required a heroic act of will. "To still your hunger, and strengthen your failing limbs, you, willing or unwilling, had to apply day after day, to the rice and water, and the half loaf of cornbread; there was no escape from it. . . . I don't remember among civilized nations any doctrine or practice of gradually starving prisoners."

6

Miss Van Lew's Spy Ring

"Employ . . . only those you know to be faithful, brave, and true."

—Union general Benjamin Butler

THE SECRET ROOM, 1863

On tiptoe, Annie followed Elizabeth Van Lew to the second floor of the family's Church Hill mansion. The young girl did not understand why her aunt was carrying a plate of food upstairs. All of the house-guests were downstairs; Annie had seen them in the dining room.

Tingling with curiosity, Annie tiptoed to the second floor, reaching it just in time to see her aunt climbing the final steps to the third floor. She was nowhere to be seen when Annie got there. Hearing footsteps overhead in the attic, Annie ascended the last flight of stairs. Her aunt's back was to her when she reached the attic; she had put down the plate and was moving a box next to the wall.

Suddenly, a panel in the wall slid open and a man's head appeared. His eyes widened when he saw Annie standing behind Van Lew, who was handing him the food. Before the stranger uttered a word to her aunt, though, Annie raised a warning finger to her lips and sought out a hiding place.

When her aunt had gone downstairs, Annie approached the man in the wall and struck up a conversation.

He was a Yankee, and he had recently escaped from Libby Prison.

* * *

As a young woman, Elizabeth Van Lew, with her blue eyes, high cheek-bones, and shrewd intelligence, was a Southern belle, the kind that wed at an early age and thence managed an affluent, bustling household. But for reasons unknown, Van Lew, at the age of forty-five in 1863, remained unmarried. In middle age, she was a birdlike woman whose quick eyes missed little.

She was the oldest daughter of transplanted Northerners, John Van Lew of Long Island, New York, and Eliza Louise Baker, whose father, Hilary Baker, was the mayor of Philadelphia when the 1798 yellow fever epidemic claimed his life. Orphaned when her mother died in 1808, Eliza was sent by relatives to live with a brother in Richmond. There she met John Van Lew, owner of a thriving hardware business, the first in Richmond. They married in January 1818 in St. John's Episcopal Church, the place where Patrick Henry had declared, "Give me liberty or give me death!" Elizabeth was born in October.

In 1836, John Van Lew purchased the three-story, fourteen-room mansion of former Richmond mayor John Adams on the summit of Church Hill. In this grand house, whose 18-foot-wide hallway stretched from front to rear, Elizabeth and her mother Eliza would live out their lives. The home was on East Grace Street, opposite St. John's, which had inspired Church Hill's name. The Van Lew home's piazza over-looked the James River and the city's warehouse district. It was one of Richmond's jewels, a place where prominent people gathered to discuss culture, ideas, and books. Guests included Supreme Court chief justice John Marshall; the Swedish author Fredrika Bremer, who wrote about the Van Lew home in her *Homes of the New World*; the world-renowned soprano Jenny Lind, the "Swedish Nightingale"; and Edgar Allan Poe, who reportedly read his new poem, "The Raven," in the Van Lew parlor.

Slavery was so tightly stitched into the fabric of Richmond society that young Elizabeth Van Lew, whose family owned fifteen slaves, at first did not question it. But her attitude began to change when, as a young girl, she was sent to the boarding school that her mother had once attended in Philadelphia. It was during her years at this school that Elizabeth fell under the influence of either a governess or a teacher who advocated abolition. She became convinced that slavery was im-moral. When Elizabeth was a teenager, she met a slave trader's daugh-ter at a Virginia spa and saw the "peculiar institution" in all of its monstrous cruelty. The girl's father had sold a female slave and her child to separate buyers. It was a common practice, but the broken-hearted mother had reportedly died of grief. The story disturbed the sensitive Elizabeth, and she grew increasingly uncomfortable with her family's possession of slaves.

As a young adult, she joined the American Colonization Society, a strange alliance of idealistic anti-slavery advocates and cynical slave-holders who wished to send troublemaking free blacks back to Africa. Later in life, Van Lew became a full-fledged abolitionist. "Slave power is arrogant—is jealous, and intrusive—is cruel—is despotic," she wrote in her diary.

Her mother, Eliza, evidently believed so, too, because John Van Lew's will, read after his death in 1843, expressly forbade her to sell or free the family slaves. The stipulation prevented Eliza from acting im-mediately on her moral principles, and jeopardizing her social status and financial security. In her meticulously researched biography of Elizabeth Van Lew, Elizabeth Varon wrote that Eliza initially honored her late husband's wishes; the 1850 Richmond tax roll listed twenty-one Van Lew slaves, fourteen in Richmond and seven at the family's outlying farm.

But the 1860 tax record cites just two Van Lew slaves—evidence, Varon suggests, that Eliza and Elizabeth skirted the will's prohibition against legal manumission by making private arrangements. Some of the slaves became family servants; others hired themselves out for wages. The mother and daughter sent a black servant child, Mary Jane Richards, to be educated in Princeton, New Jersey. When she com-pleted her schooling, Richards sailed to Liberia, the American Colo-nization Society's West African settlement for repatriated blacks, and lived there four years before returning to Richmond.

* * *

Although they were pro-Union and anti-slavery, when the war began Elizabeth and Eliza elected to remain in the mansion on East Grace Street rather than go into exile in the North. This decision meant a self-imposed sentence of censorship, ostracism, isolation, and peril. "One day I could speak for my country, the next was threatened with death," Elizabeth Van Lew wrote in her diary. Now a mature, energetic woman who lived by her convictions, she could not idly watch her neighbors and Southern countrymen wage war against her beloved Union. The sight of the Rebel flag flying for the first time over Richmond, followed by a torchlight parade, inspired Elizabeth to embrace dissidence as a lifestyle. "I never did remember a feeling of more calm determination and high resolve for endurance . . . than at that moment," she wrote. But it was more than dangerous to be seen aiding the Yankees in the Rebel capital: it was a hanging offense.

In August 1861, Elizabeth began bringing food, clothing, books, and blankets to the Union war captives held in the city's warehouses. Even such small charitable acts provoked her countrymen to mutter that she was committing treason. But she would not be deterred. When Abraham Lincoln's rabidly anti-Union brother-in-law, Lieutenant David Todd, curtly turned her away from the Ligon warehouse prison, Elizabeth Van Lew sought the help of her old friend, Confederate treasury secretary Christopher Memminger. His quiet influence helped break down the official barriers to her philanthropy. He sent her to the commandant of Richmond's prisons, General John Winder. Van Lew took care to compliment him on his white mane of hair, and Winder permitted her to not only visit the prisoners, but to send them "books, luxuries, delicacies and what she may please." Whether Winder had acted on Memminger's instructions, or because of Elizabeth's flattery, she achieved her purpose.

Every day or so, Elizabeth Van Lew or a servant would drive down Grace Street from Church Hill to one of the warehouse prisons and, after first bribing the prison authorities, would give the captives food, clothing, bedding, furniture, and books that Elizabeth and her mother had purchased. When fourteen Union officers were transferred to the

Henrico County jail—to be held as hostages for fourteen Southern privateers facing execution in the North—Elizabeth made sure that warm rolls awaited them in their cell when they arrived. Until they were sent back to Libby Prison, she brought the officers food and books. With a combination of charm, flattery, and bribery, she usually got her way with the prison authorities, and even convinced them to transfer ill inmates to civilian hospitals.

Once, when the Rebels refused to deliver her custard to sick prisoners, she made a gift of it to an influential War Department official. He sent her a thank-you note. "The custard was very nice," he wrote, adding that he had eaten it with crackers, in "fine style." On the other side of the note, Van Lew scrawled, "God help us."

Throughout 1861 and in early 1862 Van Lew smuggled letters and messages in and out of the prisons in a false-bottom plate warmer. Eventually the guards became suspicious, and a warning reached her that they planned to inspect the warmer during her next visit. Before leaving home for the prison, she filled the warmer with boiling water, and slipped it inside a cloth holder. As expected, a guard demanded to see it. Van Lew obliged, handing him the warmer while removing it from its covering. The guard yelped in pain when his bare hands came into contact with the scalding metal. Van Lew's plate warmer was not inspected again.

Van Lew's kindnesses sustained many Union War prisoners. After Private Lewis Francis had recovered from the amputation of one of his legs in a prison hospital, he wrote, "I should have perished from want, but a lady named Van Lew sent her slave every other day with food, and supplied me with clothing until January." Alfred Ely, the war prisoner and congressman from New York, persuaded Confederate officials to transfer Calvin Huson, a friend and fellow civilian prisoner who was wasting away from typhoid fever, to a place where he stood a chance of recovering: the Van Lew home. Despite the best efforts of Elizabeth and Eliza, Huson died a month later. They buried him in Hollywood Cemetery and decorated his grave with roses—a gesture that stoked the resentment of their Richmond neighbors.

In an attempt to allay suspicions about their loyalty to the Southern cause, the Van Lews took in a Confederate officer and his family as boarders; the officer was in charge of one of the Union prisons. Meanwhile, Eliza Van Lew, and her son, John, made a show of ministering to hungry, sick, and wounded Confederate soldiers. "We have to be watchful and circumspect—wise as serpents—and harmless as doves, for truly the lions are seeking to devour us," Elizabeth wrote in a diary entry in June 1862.

*   *   *

By that time, the Van Lews belonged to a Unionist underground movement that was becoming more active by the month. There were dozens of Richmond residents like the Van Lews—and hundreds, if not thousands, more who silently prayed for a Union victory, yet dared not display their subversive loyalties. But no more than a couple of dozen of them were active at any given time. For the greater part, they were middle class or well-to-do. The Unionist underground's membership changed often, although there were several others besides the Van Lews who served throughout the war. John H. Quarles and his wife; William Rowley, a farmer; Arnold B. Holmes; Abby Green; F. W. E. Lohmann, who was a restaurant owner; Franklin Stearns, a whiskey distiller; and Samuel Ruth, superintendent of the Richmond, Fredericksburg & Potomac Railroad, were prominent in the underground from 1862 until the war's end. A dozen others joined later as the movement grew more active.

The Unionists were egalitarian; the Van Lews and the other well-off members hired lawyers to represent their less wealthy compatriots when they were accused of disloyalty and gave them money when they needed to flee. Blacks such as Oliver Lewis, James and Peter Roane, and Mary Jane Richards served as message couriers and errand runners, because they could roam the city without attracting attention. In the underground movement's infancy, the Unionists simply collected intelligence and shared it with one another and the Union prisoners, whom they tried to aid when they could. They mostly communicated by letter, but sometimes they felt the need to meet face-to-face, if only to bolster one another's morale and to speak openly of their loyalty to the Union. In her journal, Van Lew melodramatically described the gatherings: "When the cold wind would blow on the darkest & stormiest night, Union people would visit one another . . . curtains pinned together . . . startled at the barking of a dog . . . the pallor coming over our faces & the blood rushing to our hearts." But sometimes the dissidents did little more than trace the Union Army's progress on maps.

As the Unionist underground matured and its members grew more confident in their ability to avoid arrest, it began aiding blacks and other refugees fleeing to the North and hiding escaped Union prisoners. The Unionists created a chain of "safe houses" inside and outside Richmond where the fugitives could find a meal and a bed, cash and passports, and where they were sometimes even outfitted in disguises and given false identification. Elizabeth Van Lew helped her black butler, William Sewell, and his family flee to the North.

To learn about impending escapes and to inform prisoners where to turn after they broke out, the Unionists made contact with the black Union captives who performed the menial jobs in the warehouse prisons. It was a relatively simple matter, because they were in daily communication with local servants and slaves who delivered goods to the prisons, and the Rebels generally ignored them. In Libby Prison, the vital link was Robert Ford, a Union Army teamster captured in May 1862; he was Richard Turner's hostler. Ford exchanged information with a black servant of Abby Green, a key Unionist who had moved to Richmond from New Hampshire before the war. Green's servant provided smuggled letters and information about troop locations around Richmond, in exchange for news of impending escapes and other developments inside Libby Prison. Ironically, the Confederates' disdain for blacks made them vulnerable to a critical security breach that might have been easily prevented.

Elizabeth Van Lew adopted the habit of taking a Sunday stroll down Church Hill and passing by the warehouse prisons. The walks were often timed so that when Van Lew came into sight, some of the black prisoners were in the captives' living quarters scrubbing the floors. They pointed her out to the white captives and told them, "That is Miss Van Lew. She will be a friend if you can escape." Lieutenant David Parker, an aide to General Ulysses Grant, noted that Van Lew kept "two or three bright, sharp colored men on the watch near Libby Prison who were always ready to conduct an escaped prisoner to a place of safety." The safe harbor might have been the Church Hill mansion, where a fugitive could hide for a day or a week in the secret attic room, or one of Richmond's other Unionist havens. When the Confederates stopped actively searching for an escapee, the fugitive would slip through the defenses ringing the city to one of the safe houses outside Richmond, and from there, strike out for the Union lines.

One of the first known Van Lew–abetted escapes occurred in 1862, when *New York Times* reporter William Henry Hurlbert, arrested while interviewing Confederate leaders throughout the South, vanished from one of the Richmond warehouses. After safely reaching the North, Hurlbert wrote a series of articles about the growing tyranny of Jefferson Davis's government. While Hurlbert never wrote about his breakout, biographer Elizabeth Varon reported that Van Lew helped him escape. About the same time, Colonel Adolphus Adler bribed a guard and walked out of Castle Godwin, a former slave jail where political prisoners and Yankees were held, in civilian clothing, hiding for a week with various Unionists before leaving Richmond. Van Lew's home, declared Adler, was "the principle place of refuge" for Union captives in Richmond.

Because she was so ready to aid escapees and practiced in helping them get out of Richmond, Elizabeth Van Lew became the underground's "guiding spirit" in assisting Yankee fugitives.

✳   ✳   ✳

In March 1862, just as Libby Prison was opening, Jefferson Davis imposed martial law in Richmond and began rooting out Union sympathizers. General John Winder's "plug-ugly" detectives arrested twenty-seven Unionists, the most prominent of them being fifty-nine-year-old John Minor Botts. As a Whig congressman, Botts had helped forge the 1850 Compromise that delayed war for a decade. Arrested in his bed in the middle of the night, Botts spent two months in a Richmond jail before he was released under house arrest to his northern Virginia farm. There, he discovered that Confederates had raided his garden and cornfields, stolen his horses, and killed dozens of hogs. A second arrest, again resulting in no charge, inspired Botts to write a rancorous letter to the *Daily Richmond Examiner*. "If to prefer living as I did before the war to living as I have done since the war makes me a traitor, then a traitor's life let me live, or a traitor's death let me die." Botts's reckless defiance helps explain his portrait's enshrinement in the Van Lew home.

The crackdown had no sooner ended than Richmond witnessed the first hanging of an American spy since Nathan Hale's execution by the British during the Revolutionary War. The condemned man was Timothy Webster, Union spy chief Allan Pinkerton's top operative. He had been hired by Pinkerton in 1853 from the New York City police force. In 1861, Webster had infiltrated a pro-Confederate Baltimore secret society that was plotting to kill President Lincoln and had foiled the assassination attempt. Three times in 1861 and 1862 he had posed as a Rebel-sympathizer courier and traveled from Baltimore to Richmond to gather detailed intelligence on Richmond's defenses, troop strengths, and batteries. But on his fourth mission, Webster was overcome by severe rheumatism, and spent weeks confined to a hotel-room bed, cared for by fellow agents Hattie Lawton and her servant, John Scobell. Concerned about Webster's long absence, Pinkerton sent two more agents to Richmond to check on him. They were captured and threatened with execution if they did not give the Rebels information. In exchange for their freedom, the agents told the Confederates where to find Webster. He went to the gallows on April 28, 1862.

Winder's crackdown and Webster's execution drove the Unionists further underground. They became craftier and more settled in their purpose. Once content to paint slogans in the dead of night on the sides of buildings—"Union Men to the Rescue!" "Now is the time to rally around the Old Flag!" and "God bless the Stars and Stripes!"—they now adopted countersigns and code names and sent each other messages written in code or with invisible ink.

✷ ✷ ✷

The Van Lews' generosity toward the Yankee captives was curtailed by martial law, but they had already been marked as women of questionable loyalty. On July 31, 1861, the *Richmond Dispatch* had singled them out in an editorial: "Two ladies, mother and daughter, living on Church Hill, have lately attracted public notice by their assiduous attentions to the Yankee prisoners confined in this City whilst every true woman in this community has been busy making articles of comfort or necessity for our troops. . . . The course of these two females . . . cannot but be regarded as evidence [of] an endorsement of the cause and conduct of these Northern Vandals."

Even after this attack, the Van Lews escaped Winder's dragnet while lesser transgressors did not. Biographer Varon believes they were shielded from arrest and prosecution by their gender, their aura of respectability, and their high social standing. Gender alone, however, was no guarantee against arrest; the Confederates sometimes imprisoned suspected women Unionists, but they were almost invariably from the working and trade classes.

While the Van Lews grew more cautious, they continued to perform acts of defiance small and large. When Union general George McClellan marched up the peninsula and briefly threatened Richmond in June 1862, Eliza Van Lew went so far as to ready a room for the general in the Church Hill mansion. "Mother had a charming chamber, with new matting and pretty curtains, all prepared for Genl. McClellan, and for a long time we called [it] Genl. McClellan's room," wrote her daughter Elizabeth.

Elizabeth Van Lew stopped attending church services when her preacher exhorted worshipers to pray for the Confederate Congress. In her journal, she wrote, "This I could not do." When the government began seizing citizens' horses wherever they found them, Elizabeth removed hers from sight. Sometimes she hid him in the smokehouse, and other times, for days on end, in her study, after spreading straw on the floor. The horse, she reported, "behaved as though he thoroughly understood matters, never stamping loud enough to be heard, no neighing." And on Jefferson Davis's designated "fast days" of self-denial and prayer, "we always tried to have a little better dinner than usual," she wrote.

* * *

One day in 1863, Erasmus Ross, nicknamed "Little Ross" by Libby Prison's Union officers because of his size, swept into one of Libby's rooms to conduct one of the twice-daily head counts. He was accompanied by the stocky prison adjutant Lieutenant John Latouche and the usual cadre of guards that policed the officers as they stood in ranks. Like the prisoners, the guards wore a medley of blue and gray uniform articles, having exchanged food and their worn-out Rebel attire for the captives' newer clothing. In contrast to the dreary monochrome attire of the guards and prisoners, Ross was a peacock. The prisoners enjoyed seeing him in his new outfits as much as Ross took pleasure in wearing them. When the head count went smoothly, Ross liked to chat and joke with the prisoners; when it went badly, he would erupt in loud profanity.

On this day, Ross strode up to Captain William Lownsbury of the 74th New York and, without warning, struck him in the stomach. "You blue-bellied Yankee," Ross sneered, "come down to my office. I have a matter to settle with you." When Ross and his Rebel guard entourage left the room, Lownsbury's comrades entreated him not to go to Ross's office; other prisoners called out by Ross had never returned, they reminded him.

But Lownsbury's curiosity was aroused, and he also knew that he could be punished if he disobeyed Ross's command. And so he presented himself to Ross, expecting a reprimand for an unknown offense. Instead, the clerk wordlessly pointed behind the counter and walked out of the office, leaving Lownsbury alone in the room.

Ross's actions puzzled Lownsbury; he wasn't sure what he was supposed to do. At last, after some hesitation, he mustered the nerve to walk around the counter and there, to his surprise, he found a Confederate uniform. Lownsbury put it on.

Dizzied by his serendipitous change of fortune, Lownsbury walked out of Libby Prison dressed in Confederate gray. No one challenged him.

When Lownsbury reached the street, a black man suddenly appeared at his elbow, signaling to Lownsbury to follow him. Up Church Hill they went—to the Van Lew mansion on East Grace Street, where Elizabeth Van Lew gave the officer directions for getting out of Richmond and finding the Union lines. Hours later, Lownsbury passed through Richmond's defenses and set out down the Virginia peninsula for the Union lines. He arrived safely.

Erasmus Ross had not simply acted on a whim in aiding Lownsbury's escape, nor was it coincidental that other captives whom Ross had summoned to his office had also disappeared. Ross was the nephew of the Unionist distiller, Franklin Stearns, and he was a deep-cover agent for Richmond's underground. His position in Libby Prison was a testament to Elizabeth Van Lew's connections in high places in the Confederacy.

By the fall of 1863, the Richmond Unionists had excellent contacts deep within the Confederate government. The underground had come a long way since the days when it scrawled pro-Union graffiti on buildings. It now was beginning to send valuable intelligence about the Confederate Army through its system of couriers and safe houses to the Union command.

Elizabeth Van Lew was as important to the Unionists' intelligence gathering as she was in aiding escapees, fugitives, and refugees. In large part due to her efforts, the Unionists had infiltrated the Rebel staffs in the city's war prisons and the War Department. She collaborated with a government engineer who made maps of the Rebel defenses around Richmond and Petersburg that Van Lew then smuggled to the Union command.

But the Unionists' pipeline to the Union Army flowed erratically, so much so that Van Lew took steps to establish a more reliable connection.

* * *

One day in December 1863, while on an apparent errand of mercy at the Libby hospital, fifteen-year-old Josephine Holmes handed a sack of tobacco to John R. McCullough, an assistant surgeon from the 1st Wisconsin who was being treated for a minor illness. Sifting through the tobacco, McCullough found a cryptic note: "Would you be free? Then be prepared to act." Josephine, whose father, Arnold B. Holmes, was a Unionist leader, returned the next day and whispered a plan of escape to McCullough. A few days later, McCullough unexpectedly "died." His friends smeared flour on his face to produce an authentic-looking death pallor and bound his hands and feet so that when the guards took the "corpse" to the Dead House, they would not have to handle the body—and notice that it was still warm and had a pulse—but could carry him by the bindings on his hands and feet. McCullough did not remain long in the Dead House. While several prisoners distracted the guards with a sham fight, he walked out of Libby with another prisoner, Captain Harry S. Howard, who was waiting for McCullough. Josephine met them at a prearranged rendezvous and led them to her father's home. Later, they were spirited to the home of Unionist William Rowley, who hid them for ten days.

When the Confederates gave up looking for them, McCullough and Howard emerged from hiding and began hiking to the Union lines, careful to avoid the batteries, revetments, and troop encampments that encircled Richmond. It was not so difficult; the defenses were built to repel invaders, not prevent people from leaving the city. They reached Williamsburg days later.

At Fortress Monroe, McCullough requested an audience with General Benjamin Butler and, when they met, he handed him a letter from Elizabeth Van Lew. After reading it, Butler had many questions about the extensive Union underground in Richmond. He was deeply interested in beginning a correspondence with the fifth columnists in the Confederate capital, but he wanted to first get another opinion.

Butler forwarded the letter to a trusted friend, Commander Charles Boutelle of the U.S. Coast Survey Office in Washington, D.C. "I am informed by the bearer that Miss Van Lieu [*sic*] is a true Union woman as true as steel." He noted that besides writing Butler a letter, she had also sent the general a bouquet. Butler was eager to have a reliable correspondent in Richmond who would be willing to drop "an ordinary letter" in the mail addressed to "a name at the North." Did Boutelle believe it prudent to send a messenger to Van Lew with instructions for conducting such a dangerous correspondence? And would she in fact do it? "I could pay large rewards, but from what I hear of her I should prefer not to do it, as I think she would be actuated to what she does by patriotic motives only." Presumably, the Coast Survey commander advised Butler to proceed, because a month later the general was writing a letter to his "Dear Aunt," and signing it "James Ap. Jones."

Butler's letter to Elizabeth Van Lew appeared to be no more than a compendium of family news: "Mary is a great deal better. Her cough has improved. . . . Your niece Jennie sends love. . . . Mother tells me to say she had given up all hopes of meeting you, until we all meet in heaven." The letter's secret message had been written with a substance, perhaps lime juice, that would pass casual inspection, but could be read after the application of heat or another reagent. It is likely that the letter-bearer gave Van Lew instructions in reading and writing invisible ink messages. Her niece Annie later remembered that her aunt possessed a small bottle of colorless liquid, which resembled water when used as ink, but which turned black when dabbed with milk.

While prison censors now routinely heated the Union captives' letters in search of hidden messages, the public's voluminous correspondence was not rigorously screened unless it aroused suspicion. Even in that event, it would have been rare for a Southern mail official to examine a letter for evidence of invisible writing. As the relationship deepened between the Union command and the Richmond underground, they more frequently bypassed the Southern mail system altogether and entrusted their letters to couriers.

"My Dear Miss," began Butler's secret message. "The doctor [McCullough] who came through and spoke to me of the bouquet said that you would be willing to aid the Union cause by furnishing me with information if I would devise a means." The general outlined how such a correspondence would be conducted. Letters should be addressed to James Ap. Jones in Norfolk, "the letter being written as this is," and sent on the flag of truce boat that shuttled between Fortress Monroe and City Point. "I am rejoiced to hear of the strong feeling for the Union which exists in your own breast and among some of the ladies of Richmond."

* * *

In early 1864, Elizabeth Van Lew began sending Butler information painstakingly gathered by the Unionists. The information about the Confederacy's operations and plans came not only from sources within the Confederate command, but also from contacts outside the government: war captives; farmers; merchants; housewives. Among the informants were Martin Lipscomb, who supplied goods to Confederate forces and ran unsuccessfully for Richmond mayor while he was spying for the Union, and one of Winder's own detectives, Philip Cashmeyer. By communicating directly with Butler, Van Lew was placing herself at great risk. Spencer Kellogg Brown had recently been hanged as a spy at Camp Lee outside Richmond, and Timothy Webster's execution remained a vivid memory. If she were caught, Van Lew's social standing would not save her from the gallows.

At first, she merely forwarded the raw information that reached her, but before long Van Lew was synthesizing the intelligence into cogent reports. Sometimes they were written in invisible ink, addressed to James Ap. Jones in Norfolk; other times, she composed the reports in a code devised by Union intelligence officers. She kept the code key hidden inside the case of her pocket watch. It was a matrix of thirty-six letters and numbers, six down and six across, with each row and column headed by a number. She signed the reports "Babcock," and later, "Romona." Van Lew secreted the messages inside hollow bronze animal figures mounted on columns flanking her library fireplace. When servants from the Van Lews' 12-acre farm south of Richmond brought produce into the city, they collected any reports that had been hidden in the library figurines, along with maps and plans purloined from the Confederate command. After carefully concealing the messages—sometimes in the heels of their shoes, or in a hollowed-out egg in a basket of real eggs—they took them to the flag of truce boat at City Point, or to an agent from the Union's Bureau of Military Information who met them at one of the five Unionist "stage stations" along the Charles City road.

When General Ulysses Grant took command in 1864, Van Lew redirected her correspondence from Butler to General George Henry Sharpe, the Army of the Potomac's intelligence chief, and to Grant himself. She was in such close contact with Grant that sometimes flowers cut in her garden in the morning adorned the general's table at suppertime.

The correspondence was reciprocal. The Union command peppered Van Lew with questions. A note hidden in a Butler letter to his "Dear Niece" was full of requests for information: "Give what account you can of the rebel rams [ironclads and naval rams were serviced at the Confederate Naval Yard, visible from the Van Lew home]. . . . Will there be an attack in North Carolina? How many troops are there? Will Richmond be evacuated? . . . Give all possible facts." Upon receiving such a letter, Van Lew would appeal for assistance to her fellow Unionists and her spies in the Rebel government.

Butler's connection with Van Lew branched into direct correspondence with two other Richmond Unionists, Samuel Ruth, the railroad superintendent, who supplied excellent information on the transport of Rebel troops and supplies; and William Rowley, who worked closely with Van Lew but also oversaw his own network of sources and informants. The general requested and received $100,000 in Confederate funds for covert operations from Treasury Secretary Samuel P. Chase, and sent half of it to Rowley to pay for information. "Please see our friends and have in working order at once," Butler wrote in an invisible-ink message. "Employ . . . only those you know to be faithful, brave, and true." The informants did not always want to be paid in nearly worthless Confederate scrip; at different times, Van Lew requested a pair of shoes, gunpowder tea, and a "muff of the latest style."

* * *

Elizabeth Van Lew went to bed every night with her papers within arm's reach, so that she could destroy them quickly if Confederate detectives raided her home. "Written only to be burnt was the fate of almost everything which would now be of value," she acknowledged. It was a stressful life; it was no wonder that acquaintances described Van Lew as nervous and high-strung. A diarist in normal times, during the war Van Lew maintained only an "occasional journal," which she buried on her property.

Increasingly, she was threatened, followed, and watched. "Have turned to speak to a friend and found a detective at my elbow," she wrote. On another occasion at her home, "Strange faces could sometimes be seen peeping around the columns and pillars of the back portico." A group that called itself the "White Caps" left a note bearing a skull, crossbones, and a crude, semicoherent warning: "They are coming at night. Look out! Look out! Look out! Your house is going at FIRE. . . . Old Maid. Is your house insured?"

In an attempt to dispel growing suspicions about her activities, she made a show of bringing gifts and books to a South Carolina unit stationed in Richmond, but it is unlikely that the gesture persuaded anyone of her fealty to the Rebel cause. She grew reluctant to even walk past Libby Prison; when she did, she dared not look up at the windows. In her occasional journal, Van Lew wrote, "I have had men shake their fingers in my face and say terrible things. We had threats of being driven away, threats of fire, and threats of death. . . . Surely madness was upon the people!" She began wearing old, coarse clothes when she went out so that she would not be readily recognized.

In 1864 the Confederate government investigated Van Lew. Her precautions served her well, as did her gender, her social status, and her habitual secrecy, even inside her home. Under questioning, her sister-in-law, estranged from John Van Lew and disgruntled with the entire family, produced for General Winder's "plug-uglies" nothing more damning than Van Lew's expressed opinions. The investigators concluded that Van Lew had done nothing "to infirm the cause—Like most of her sex she seems to have talked freely . . . in the presence of female friends."

They shockingly underestimated her. Not only was Elizabeth Van Lew one of the Unionists' top intelligence leaders in Richmond, she also might have pulled off one of the war's greatest espionage feats: smuggling a spy into the Confederate White House to eavesdrop on the dinner conversations of Rebel leaders. While documentation is sketchy at best, Richmond legend identifies the spy as Mary Jane Richards, the black servant child whom the Van Lews sent to the North to be educated, and who lived in Liberia before returning to Richmond. Richards had married a William Bowser at St. John's Church in 1861, and a Mary Jane Bowser purportedly was a waitress in the Confederate White House. It is possible that, standing at the elbow of the Confederacy's highest-ranking officials, Elizabeth Van Lew's former servant gleaned choice information from the highest Rebel officials and passed it along to the Union underground.

* * *

In January 1864, Elizabeth Van Lew and her fellow Unionists began receiving vague reports from their Libby Prison informants that something important was going to happen at the prison. Robert Ford had apprised Abby Green's servant, who had told Green, who was in close contact with Van Lew. When better intelligence reached her—that "there was to be an exit," she wrote—Van Lew and her servants converted a room in her home into a sanctuary for fugitives. They nailed blankets over the windows, moved in cots and mattresses, and kept a small gas fire burning day and night for weeks. "We were so ready for them," she later wrote.

The information was disseminated to Unionists who might aid the fugitives. Richmond's underground prepared for whatever might come.

7

The Warrior Schoolteacher

"A prisoner, if he deserves the name, is always more or less occupied with the idea of making his escape."

—Colonel Federico Cavada, Libby Prison

LATE SEPTEMBER 1863

Colonel Thomas Ellwood Rose awakened on a hard floor inside Libby Prison, sleep-deprived and disoriented. The Pittsburgh schoolteacher ran his fingers through his dark, curly hair and bushy beard, both gray-tinged, although he was just thirty-three and his face was yet unlined. Even to a veteran soldier accustomed to change such as he, this transition from combat to captivity was a shock. Upon finally reaching Richmond, the new prisoners had been robbed and thrust into a reeking room filled with gaunt men in ragged uniforms crawling with lice. Then had come the long night on the spittle-flecked floor, with nothing to cushion their bones. At breakfast the new prisoners were introduced to the sharp-elbowed "army of ferocious cooks," hacking and red-eyed amid clouds of wood smoke at the stoves.

If it all seemed dreamlike, Rose had his throbbing foot to remind him of the grim reality of his situation. On September 19, he had led the 77th Pennsylvania through a hailstorm of lead and steel in

Chickamauga's burning woods. Hours later, he was a prisoner of the Confederates.

* * *

Born and raised in a Quaker area of eastern Pennsylvania along the Delaware River, Rose grew up as a Presbyterian and became a school-teacher like his father. In 1852, he married Lydia C. Trumbower, and they moved to western Pennsylvania, where Rose was principal of South Pittsburgh Schools. When the war began, he enlisted as a private in the 12th Pennsylvania, a Pittsburgh infantry regiment. The quiet teacher was soon promoted to captain and given command of two companies of the 77th Pennsylvania.

Rose fought at Shiloh, and eight months later at Stones River, Tennessee, he became the 77th's commander in midbattle when Colonel Peter Housum was mortally wounded. Rose led a pivotal counterattack that so impressed General William Rosecrans that he pronounced the 77th to be "the banner regiment at Stones River," and promoted Rose to colonel. Three months before Chickamauga, the 77th lost one-third of its men at Liberty Gap, Tennessee, while driving the enemy from a tactically important hill.

* * *

After Gettysburg and Vicksburg, General James Longstreet had quietly transferred 20,000 Confederate troops from General Robert E. Lee's army in Virginia to General Braxton Bragg's army near Chattanooga. With their combined force of 60,000, Longstreet and Bragg hoped to crush General Rosecrans's smaller army and end the string of catastrophic Confederate defeats.

At Chickamauga Creek in northern Georgia, the plan succeeded brilliantly. Rosecrans's broken army escaped annihilation only because of General George Thomas's heroic stand on September 20. The battle's namesake creek lived up to its Cherokee name, which means "river of death." Confederate and Union losses totaled 34,400 killed, wounded, and missing. The only battle of the war with more casualties was Gettysburg, which lasted one day longer. As Rosecrans's shattered forces tried to regroup in Chattanooga, Bragg hesitated. Rather than attack right away, Bragg placed Chattanooga under siege, squandering a magnificent opportunity to destroy Rosecrans's army.

At nightfall on September 19, the first day of the battle, two Texas regiments had burst through a gap in the Union line, flanked the 77th Pennsylvania's advanced position, and captured all of the regiment's field officers, seven line officers, and seventy men. The Pennsylvanians were held for several days before being herded into cattle cars to travel to Richmond, via Columbia, South Carolina, and Raleigh. During the journey, they were robbed, spat upon, and abused. At one point an elderly Southern woman asked the guards to shoot the captives. They also encountered unexpected kindnesses; at a South Carolina siding, Confederate troops tossed the hungry captives chunks of bread and meat.

Colonel Rose was already looking for a means of escape. In Weldon, North Carolina, he leaped from the train and ran. Rebel troops recaptured him in the soggy woods after a day's desperate freedom. Nursing a broken foot, Rose was hustled back to the train.

In Richmond, the captives marched through a gantlet of hostile Southerners lining the streets between the railroad station and Libby Prison. As the onlookers hissed at and spat on the authors of their privations, a heckler called out, "Oh, is these the kind of brutes that has come down here to kill our noble sons?" A Richmond newspaper declared that death was too good for the Yankee prisoners. "First, it gives them no time to repent of their folly and wickedness, and, second, no matter how we treat them, they will invade a territory of fire and brimstone which will make them want to come back here."

* * *

Twelve days after his capture, Colonel Rose limped painfully around the crowded second-floor room, soon to be known as the Lower Chickamauga Room because so many of its inhabitants had been captured during that battle. He made a minute examination of the window casings, fireplaces, and whitewashed walls. He began committing to memory the streets, buildings, and terrain around the prison, noting the proximity of the Kanawha Canal to the south side of the prison on Canal Street. Careful to not present a target for the guards, he steered clear of the window openings.

Rose watched with keen interest as workmen climbed into and out of a large sewer on Canal Street, concluding that the sewer must empty into the Kanawha Canal. The outlines of a plan began to take shape in his mind. Other captives noticed the quiet colonel's unusual single-mindedness. "From the hour of his coming, a means of escape became his constant and eager study," wrote Lieutenant Frank Moran.

Rose's survey of Libby's accessible rooms led him down the stairway from the first-floor kitchen to the east cellar room, the area of the prison closest to the sewer. To accommodate the new captives from Chickamauga, prison officials had set up a second kitchen in the cellar room. It was a filthy, sepulchral, rat-infested place, and the prisoners avoided it whenever possible.

It took a few minutes for Rose's eyes to adjust to the dim light before he could begin a careful inspection. As he moved into a darker area of the room, he unexpectedly bumped into someone standing in the shadows.

As a regimental commander, Rose had few confidants among his fellow Pennsylvania officers. He was wary, too, of the other prisoners, who were strangers and not yet to be trusted. The man with whom Rose had collided was just as alarmed and suspicious as Rose, but they nonetheless cautiously struck up a conversation, acutely aware of their mutual peril. Confederate informants lurked among the prisoners, and a careless remark could send a prisoner to the dungeon in the middle cellar.

Rose's wiry-thin new acquaintance was Major A. G. Hamilton of the 12th Kentucky Cavalry. Hamilton, a builder of homes in civilian life, was captured in late September during a clash with Confederate cavalry in the Appalachian Mountains near Jonesborough, Tennessee. Like Rose, he inhabited the Lower Chickamauga Room—the middle room on the second floor. Hamilton described his days in Libby as "going by like scarcely moving tears." The nights, he wrote, passed "like black blots dying out of a dream of horror."

Rose and Hamilton quickly recognized that they were in the cellar for the same purpose: to find a way out of Libby Prison. "Our acquaintance," wrote Hamilton, "ripened into a mutual friendship and we soon had the full confidence of each other." Of Libby's many "strong-minded men with courage of steel," noted Hamilton, the modest, unassuming Colonel Rose occupied the "front rank."

Rose and Hamilton agreed that they must somehow break out. Believing it foolhardy to try escaping through the prison's doors or windows, they resolved to dig out. The new partners sought a place to begin a tunnel.

The temporary kitchen's large cauldrons rested in a crude, furnace-like contraption in one corner. The stench of grease, rotten food, and sewage made them gag as they meticulously examined the room. The smell, the heavy gloom, and the low ceiling were suffocating. Underfoot scurried swarms of squealing river rats, which slipped as easily through the moist walls as if they did not exist. Libby's rats were the stuff of legend. Whenever the James River overflowed its banks and flooded the cellars, wrote one prisoner, "enormous swarms of rats come out from the lower doors and windows of the prison and . . . head for dry land in swimming platoons amid the cheers of the prisoners in the upper windows."

Rose and Hamilton paid special attention to the east wall as a possible starting point. In a recess behind a building support, they found a place where the soil was compacted enough so that it did not cave in on them.

They did not begin work right away. They first satisfied themselves that if they dug at that spot they would not be disturbed or caught in the act. Then they had to procure excavation tools. A few days passed before they were able to obtain them—two case knives and a broken shovel, scarcely ideal implements for such an ambitious project.

At last, they began to dig. The deep shadows cast by the building groin concealed their work from the cooks and captors who congregated around the cauldrons across the room. They believed that with a lot of hard work and a little luck they could dig to the sewer, break into it, and follow it to freedom.

But then, before they had advanced the tunnel very far, the Confederates abruptly transferred scores of officers to Danville, Virginia, and released all of the captive Union surgeons and chaplains. This was an attempt to resolve Libby Prison's worsening problem of overcrowding. With fewer prisoners, the Confederates saw no further need for an auxiliary kitchen, and they closed the cooking area in the east cellar. The cauldrons were removed, and the furnace that had been used to heat them was trundled outdoors and discarded. Finally, the Rebels sealed the stairway to the east cellar after two officers sawed through the cellar's wooden-barred window and escaped, reaching the Union lines a week later.

Without access to their tunnel, Rose's and Hamilton's short-lived partnership dissolved. They returned to their separate pursuits of a way out. But it seemed that wherever their explorations took them, they encountered one another. Their most memorable meeting occurred one night on a high scaffold outside the upper west room.

Workmen had left the scaffold standing against the outer west wall at the end of the day, an oversight that escaped neither the notice of Rose or Hamilton. There was a storm that night, and heavy rain normally drove the Rebel guards indoors. Rose and Hamilton separately planned to climb onto the platform, drop to the ground, and flee the prison grounds. Both men slipped through a window onto the scaffold, neither aware of the other's presence—until they bumped into one another in the dark.

Badly startled, they nearly came to blows until a flash of lightning revealed their faces to one another. Indeed, the lightning bursts came so often and with such intensity that they were forced to abandon their escape plan and return to the prison through a window.

Henceforth, they worked as a team.

* * *

NOVEMBER 1863

Rose and Hamilton remained convinced that Libby's cellar was their best hope of escape, even with the east cellar stairway now sealed. They ruled out the west cellar, where the black prisoners lived, coming and going through a guarded exterior door; directly above it were the Rebels' sleeping quarters and offices. That left the middle cellar, with its carpentry shop and jail cells for hostages and troublemaker captives. Located beneath the main-floor kitchen, the middle cellar appeared to be the only basement room that might be entered from inside Libby.

One night, Colonel Rose pried up a floorboard in the kitchen with a piece of pine he had whittled into a wedge. Through the opening, he lowered a long board wrenched from a table seat in the kitchen. When it touched the cellar floor, it extended a foot above the kitchen floor. After satisfying himself that no one was watching, Rose slid down the board into the middle cellar.

When his eyes adjusted to the darkness, Rose saw that there was an entryway onto Canal Street, and that it had no door. Hugging the wall, Rose stood so close to the Rebel sentinel pacing outside that he could have reached out and touched him. Turning his head, he noted that the jail cells were unoccupied. Rose stared at the doorway, calculating distances and odds, and picturing the layout of the buildings and guard posts. He felt a flutter of cautious optimism. After shimmying up the board to the kitchen and returning everything to its proper place, Rose held a whispered conference with Major Hamilton.

They returned to the kitchen the next night with a 100-foot length of inch-thick rope obtained from Colonel Harry White, who distributed bales of clothing to the Belle Isle prisoners. Removing the floorboard, Rose tied the rope to a supporting post. He and Hamilton slid down the rope to the cellar. Over the next several nights, the men familiarized themselves with the room and watched the Confederate guard as he paced along his narrow strand, counting the steps that he took between turns. When a second sentinel appeared, they worked out how long the doorway was out of the sight line of both guards.

* * *

A week after Rose's first trip to the middle cellar, he and Hamilton had become confident that they could dash to Canal Street and hide in the shadows without being seen. They decided to chance it. Rose went first. Darting through the cellar doorway, he was headed toward Canal Street when the second guard spotted him and shouted for the corporal of the guard. The guard chased Rose into the cellar, but he and Hamilton managed to climb the rope to the kitchen and replace the floor plank without being seen. Guards invaded the middle cellar, rousting several civilian laborers who had worked in the carpentry shop that day and were sleeping on bedrolls in a corner. It was a lucky break for the would-be escapees, for the guards concluded that they must have seen one of the workmen, and not a prisoner trying to escape.

Rose and Hamilton were so encouraged by their near success that they enlisted other prisoners—enough to overpower the guards, if necessary. Singly and in pairs, the recruits accompanied the ringleaders on nightly trips to the middle cellar to learn its layout. Days after the aborted escape attempt, all the elements were in place for a more ambitious breakout.

On the appointed night, at the appointed hour, the escape party silently climbed down the rope to the middle cellar. But no sooner had they all assembled than a prisoner lookout upstairs sounded an alarm:

guards were converging on Libby. The would-be escapees instantly began ascending the rope, one at a time. It seemed to take hours. Rose, the last to leave the cellar, had just replaced the floorboard when guards burst into the kitchen. Thinking fast, the colonel sat down at one of the tables and stuck a pipe in his mouth, affecting a nonchalance that he certainly did not feel. The guards jogged past Rose without a second glance and went upstairs, where they looked around the living quarters. Finding nothing suspicious there, they left.

Libby buzzed with rumors of an impending breakout. Prisoners beseeched Rose and Hamilton to include them in their no-longer-secret escape plot. Rose and Hamilton obligingly put the other prisoners under oath; before long, 420 men had taken the pledge.

But Rose and Hamilton now worried that there were too many conspirators. Although there were more than enough prisoners to overpower Libby's guard force, such a large breakout would bring down a host of armed militiamen. Many of the prisoners would be killed or wounded. And there was another problem: Rose and Hamilton were unsure of the location of the nearest Union lines.

With great reluctance, they announced that they were abandoning the scheme. The frenzied discussions of escape died away.

* * *

With the middle cellar now being watched and too dangerous for use as a jumping-off point, Rose's and Hamilton's private tactical discussions returned to their initial object: a tunnel from the east cellar.

The east cellar's inaccessibility made it even more desirable to them as a staging area for a breakout. Not all the captives agreed. Colonel Federico Cavada observed that the east cellar's difficulties appeared to the other prisoners to be "more impossible than they really are." This was understandable, for the difficulties loomed large indeed. The first formidable challenge was breaking into the sealed room without being seen or heard.

Rose and Hamilton ruled out the prison hospital directly above the east cellar as an access point. If they had been inclined to try it, they would first have had to gain entry to the hospital, then remove some floorboards inside the hospital so that they could descend into the cellar. It was ludicrous to imagine they could accomplish this in plain sight of patients, orderlies, doctors, and guards.

The kitchen was not an obvious portal. It was on the first floor of the middle warehouse, and the east cellar was a level below it, in the adjacent building. From the kitchen, Rose and Hamilton would have to descend to the carpentry shop in the middle cellar, and then cut a hole in the brick wall separating it from the east cellar. This plan's fatal flaw was that the workmen who used the middle cellar's shop during the daytime would notice the sudden appearance of a hole in the cellar wall.

It was probably Hamilton, the Kentucky house-builder, who recognized the kitchen fireplace's possibilities as a passageway to the east cellar. Because its firewall separated the kitchen from the hospital, any excavation would have to be shallow, lest they breach the hospital's wall. Hamilton would hollow out a tiny room within the firewall, then dig down 4 or 5 feet—safely below the level of the hospital floor. At that point, he could break through the east cellar wall, being careful to preserve the adjacent carpentry shop wall. The excavation would resemble an inverted S, its top being the kitchen fireplace, and its bottom the east cellar. After carefully studying the plan, Rose became convinced that Hamilton, with his skills and sure touch, could carry it off.

From reveille to lights-out, the kitchen was a busy place, where captives prepared food, ate, exercised, danced, and staged concerts. The Rebels had moved most of the kitchen's ramshackle stoves to the upstairs rooms so that the cooks could prepare meals there. But some of the captives preferred cooking on the two remaining kitchen stoves—which they heated with kindling stacked in the fireplace—to the rough-and-tumble of the upstairs rooms.

With lights out at 9 PM, the kitchen fell dark and silent, and Rose and Hamilton could then wedge themselves behind the cookstoves and scrape and dig until 4 AM, when the early risers began trickling into the kitchen to light a fire. If everything went according to plan, it could work. But a sleepless prisoner entering the kitchen at an awkward moment, or Rebel guards making an unscheduled inspection could just as easily expose their scheme.

* * *

On December 19, after the upstairs rooms had fallen silent except for the snores and muttering of sleeping men, Hamilton and Rose quietly rose from their sleeping places and crept into the kitchen. They carefully moved aside the two stoves in the fireplace, meticulously collecting the soot and ashes in a rubber blanket. With a borrowed jackknife, Hamilton began digging out the mortar between the bricks at the rear of the fireplace. The work advanced slowly because of the utter darkness and the crucial need for silence.

When they heard sentries at Libby and nearby Castle Thunder cry the hour of four o'clock, Hamilton and Rose carefully replaced the bricks that Hamilton had painstakingly pried loose and removed while Rose stood lookout all night. Hamilton's hands were scratched and dirty from the rough work. The two men flung handfuls of the soot that they had collected in the blanket against the firewall. In the places where Hamilton had removed the mortar, they carefully filled the seams between the bricks. After returning the stoves to their usual places, Rose and Hamilton tiptoed upstairs to snatch a couple of hours of sleep before reveille.

The excavation became a punishing routine. The officers would noiselessly slip into the kitchen, move the stoves, and sweep the fireplace ashes and soot into the rubber blanket. Hamilton worked patiently, and Rose sometimes traded places with him, chipping at the wall with the jackknife or a chisel that he had stolen from the carpentry workshop during a nighttime visit. The men observed their strict protocol of laboring without a light and in near absolute silence. Just 10 feet away, on the other side of the prison wall, a sentinel paced back and forth.

They removed the mortar cementing fifty to seventy-five fireplace bricks and made a hole large enough to admit a man inside the wall between the kitchen and the hospital. Hamilton then began digging downward. He passed what seemed like an eternity of nights in the inky-black chamber before descending below the level of the hospital floor.

On December 30, eleven days after they had begun the excavation, Hamilton broke through the wall into the east cellar.

They could hardly be blamed for pausing to savor their secret triumph. Not only had they reached the east cellar without being detected, they had done it with nothing but a pocketknife and a chisel. "It seemed as though half the battle had been won," wrote Hamilton, "although in reality our labors were barely commenced."

* * *

The completed passageway was narrow and hazardous, and on one of his first trips through it, Rose nearly died. He entered the chute backward, on his hands and knees, feet first. Not yet accustomed to bracing himself against the walls with his back, hands, and feet, he slid all the way to the first bend in the passage and became wedged with his arms pinned to his sides. The more he struggled to free himself, the more firmly lodged he became. His neck and back were bent at such a sharp angle that he could hardly draw a breath. As Rose gasped for air and began to lose consciousness, Hamilton vainly tried to wrench him free.

Hamilton had to find help, or Rose would suffocate in the fireplace. Spontaneously enlisting a new ally in the middle of the night was supremely risky, but there was no time for cautious half measures. He raced upstairs to the Lower Chickamauga Room, and frantically hunted for Major George Fitzsimmons of the 30th Indiana, a trusted friend. After stepping all over the sleeping men on the floor, "leaving riot and blasphemy in his track," as one observer noted, Hamilton could not locate Fitzsimmons among the rows of sleepers. Desperate now, Hamilton impulsively awakened Lieutenant F. F. Bennett of the 18th U.S. Regulars, and led Bennett to the kitchen. Together they were able to rescue Rose from asphyxiation.

Clearly, if the fireplace was to be their portal, it needed improvement. Hamilton set about enlarging the opening with the jackknife and chisel, while Rose canvassed the prison for a coil of rope to ease the transit from the kitchen to the cellar. The rope and Hamilton's renovations made the descents many degrees easier.

In the cryptlike east cellar, amid the squealing river rats, Hamilton and Rose revived their plan for a tunnel to the sewer. Rose believed that if they could dig to the sewer, they could easily walk through it to its outlet at the Lynchburg Canal yards away.

After two attempts to breach the east foundation wall ended in cave-ins, on the third attempt Rose and Hamilton found a spot in the southeast corner where the soil was more tightly packed. They began digging a tunnel.

* * *

Incredibly, despite Rose's and Hamilton's twelve nights of hewing a corridor through brick and mortar, and Rose's near fatal sequestration in the wall, fewer than six men among Libby's twelve hundred captives knew about the passageway in the kitchen fireplace. Hundreds of men visited the kitchen daily, ignorant of the fireplace's great secret.

Rose and Hamilton each night rose from their sleeping places, stole into the kitchen, and slid down the rope into the east cellar. After Hamilton had finished dilating the chute, they cached their primitive digging tools—now two knives, the chisel, and a broken shovel—in the cellar, lest they be caught with them during a Confederate raid of the upstairs rooms for contraband.

The east cellar reeked of rancid pork fat, wet clay, sewage, and the penetrating, feral odor of rats. Pale light from the Canal Street lamps gleamed through gaps around the door and through the barred window, though it did not dispel the room's thick, cloying gloom.

The rats were chronic nuisances. They scurried over the men's feet when they were standing, and walked over their hands and legs when they dropped to their knees to dig. But Rose and Hamilton had largely overcome their feelings of revulsion during their earlier excavation. Ignoring the rodents and the stench, they labored without interruption.

Already knowing the lay of the place, they quickly sketched a plan of action. After first digging down and under the east wall, they would turn south toward the 6-foot-high sewer, less than 20 feet away, that emptied into the canal. Once they breached the pipe, they could travel inside it to the canal and vanish into the darkness.

*　*　*

Rose and Hamilton were practical, efficient men. Hamilton had utilized his mastery of the building trades to systematically disassemble and carve a secret passageway inside a brick-and-masonry fireplace. It was now Rose's turn to display his rare talent. As it turned out, he was an extraordinarily industrious tunneler, undaunted by nighttime labor inside a claustrophobic, 2-foot-diameter burrow. Rose's unusual skill as a digger and his incomplete personal history suggest that, between teaching jobs, he might have worked in one of the many Pittsburgh-area coal mines.

The system that Rose and Hamilton devised was simplicity itself: Rose dug, and Hamilton did everything else. Lying on his stomach or back in the cramped work space—at first with a candle, but later, when the tunnel lengthened and oxygen became so scarce that a candle would not stay lit, in tomblike darkness—Rose inched forward like a mole, burrowing with the chisel, the jackknife, and his bare hands when the packed earth had been pulverized into loose dirt. He never panicked, even during the times when reefs of dirt gave way and half buried him, filling his nose and mouth. With his hands and the broken shovel (also effective for fending off rats), Rose shoved the loose dirt into a wooden spittoon taken from an upstairs room. When it was full, Rose tugged on the clothesline fastened to the spittoon's base, the signal for Hamilton to drag it from the tunnel, empty it, and return the spittoon to Rose. Hamilton hid the excavated dirt under piles of discarded straw from hospital mattresses and relief boxes.

With his hat, Hamilton fanned the tunnel entrance to push fresher air into the tunnel, but less of it reached Rose the farther he burrowed from the tunnel portal. They tied a rope to Rose's foot in case he fainted and had to be dragged out. Rose worked in spurts, sweating heavily and panting, and then backing out of the hole to recover in the cellar's dim light and fetid air, which seemed blessedly pure compared to the dirt grave.

After a few days of this work, Rose and Hamilton recognized that they needed helpers. It was impossible for one man alone to fan air into the tunnel, empty the cuspidor, and hide the dirt, while also standing lookout. After analyzing the components of the tunnel-digging operation, they decided to create a much larger workforce, and to organize the men into shifts. With a fresh team at work each night, the tunnel could be completed all the more quickly.

Still, each additional man sworn to secrecy heightened the risk of exposure, which meant banishment to the dungeon. It was a gamble. Some of the prisoners, nicknamed "toadies," tried to ingratiate themselves with their captors, sometimes by acting as informants in the hope of being exchanged first. Far more dangerous were the spies planted in the prison by the Confederates. Disguised as war prisoners, they mingled freely with the captives, eavesdropping and sometimes initiating discussions of escape and insurrection. Rose and Hamilton feared the spies above all.

✶ ✶ ✶

January 1864

After carefully screening candidates for three digging teams, Rose chose thirteen trustworthy men and "bound them by a solemn oath to secrecy and strict obedience." One of them was Lieutenant Bennett, who had helped rescue Rose from suffocation. Rose and Hamilton introduced the new men to the secret passageway, and to the workings they had begun in the east cellar. With the fireplace chute now more heavily traveled than ever, one of the new tunnelers, Major B. B. McDonald of the 101st Ohio, transformed the climbing rope into a more easily scalable rope ladder.

Rose formed three five-man squads; each man was given a task that suited him—digger, spittoon emptier, fanner, relief emptier and fanner, or lookout—but they served in other roles as needed. After a night's work in the cellar, each squad rested two nights, while watching the Rebel sentinels from the windows for signs of an impending surprise raid.

When their initial enthusiasm faded, some of the recruits found that they could not tolerate working in the cellar. "The indescribably bad odor and impure atmosphere of the cellar made some of them sick," wrote Rose. Others could not abide the ubiquitous rats or the prolonged contortions forced on them by the work. Everyone initially was disoriented by the cellar's profound darkness, and, because silence had to be observed, the men at first had difficulty even finding the tunnel. Rose sometimes had to gather up lost men by groping around the cellar in the dark.

Only Rose and Hamilton labored consistently and efficiently in the nauseating, vermin-ridden environment. Lieutenant Frank Moran of the 73rd New York, who was not one of the tunnelers but profited from their labor, described Rose as "by long odds, the best digger of the party, while Hamilton had no equal for ingenious mechanical skill in contriving helpful little devices to overcome or lessen the difficulties." One modest innovation was replacing Hamilton's much-used hat for fanning air into the tunnel with a rubber blanket stretched over a light frame.

Their unshakable faith in their project inspired Rose and Hamilton to persevere, while their comrades were more vulnerable to private doubts. Even Rose conceded that "to the unreflecting the scheme seemed impracticable as soon as the first burst of enthusiasm was over."

* * *

At first, the digging went relatively smoothly, once the team members grew accustomed to working together and the cellar's darkness no longer seemed so daunting. They obtained another jackknife and some case knives, although the chisel and bare hands remained the best tools for digging through the red clay.

But the modest improvements mitigated neither the crews' exhaustion and boredom, nor the brutish nature of the work. At the end of each shift in the cellar, the diggers returned wearily to their spots on the floor—filthy, reeking, and utterly drained, their shoulder muscles burning, backs aching, and hands, knees, and elbows skinned and bruised. If they had the energy to conjure up one thought before falling asleep, it was gratitude for the two nights of rest stretching ahead of them before they had to re-enter the cellar and work another shift.

The crews were advancing the tunnel by yards each night when they encountered their first major obstacle: the ponderous, foot-thick timbers that supported Libby's east warehouse. Rose's workers did not see how they could cut through the timbers with their puny knives and chisel. A few were ready to give up.

Someone, very possibly the inventive Hamilton, proposed remaking the old knives into miniature saws, with teeth. When this was done, the prisoners attacked the timbers again. After "infinite labor," the thick wood succumbed to thousands of cuts, and the tunnelers were digging in red clay again. The men's spirits rose. Just yards away lay the sewer, and liberty.

While burrowing through the clay one night, Rose encountered a rivulet of water. He tried to ignore it, but the water flow increased as he forged ahead. And then, without warning, Rose was struck with great force by a torrent that flooded his cramped work space from floor to ceiling. Rose's life was saved by the routine precaution that he and Hamilton had adopted of tying a rope to the tunneler's foot. The other workers dragged the colonel, gasping and coughing, from the tunnel.

Later, Rose and Hamilton concluded that in tunneling downward from the cellar to the sewer, they had made the angle of descent too steep and had dug below the level of the Kanawha Canal.

They sealed the tunnel, careful to leave no trace of their work there. It was of no further use to them.

But they were not ready to quit.

8

Test of Faith

"The feelings of that little band, who can describe! From hopes almost as bright as reality, they were suddenly plunged into the depths of despair."

—Captain I. N. Johnston

LIBBY PRISON, JANUARY 1864

Despite the previous tunnel's failure, Rose firmly believed that the sewer remained the most promising escape route. His careful survey of the prison grounds had yielded a new insight: there was a smaller sewer near the east cellar that emptied into the large sewer. A tunnel to the small sewer might enable them to utilize it to reach the large sewer, and then walk through the large sewer to the canal bank and freedom. Rose rounded up his three teams and laid out his new plan.

The captives began the new excavation near the flooded tunnel in the cellar's southeast corner, taking care that they did not burrow downward too far, lest they admit water from the Kanawha Canal. They stole a hatchet, an auger, and a piece of iron bar from a Rebel work crew that was replacing Libby's wooden window bars with ones made of iron. "They [the Confederates] were industriously engaged in making . . . the prison proof against escape, yet . . . they were affording us at the same time, the opportunity of procuring the very identical tools with which we *did* make our escape," wrote Captain David Caldwell.

If it hadn't been for the choking stench of raw sewage that worsened as they inched toward the small sewer, they would have been grateful for the east cellar's relative warmth. The captives shivered in their tattered summer-issue uniforms as unusually cold January winds and snow gusted through the upstairs rooms. On January 4, four inches of snow fell and, days later, another inch accumulated. The temperature did not climb above freezing. The prisoners watched ice skaters glide over the James River where it began its turn southward, and perhaps dreamed of Northern winters gone by, or of warm summer days.

They had been at work for several days when something unanticipated occurred: the tunnel abruptly caved in. After a swift examination of the site from inside the tunnel and from an upstairs window, Rose and Hamilton determined that they had dug beneath a castoff brick furnace resting on the ground outside the prison wall. Once used in the east cellar's temporary kitchen, the furnace had been carted outdoors and left there when the kitchen was closed. The furnace's massive weight had crushed the tunnel where it passed beneath the brick ruin.

Of course, the Confederate sentinels, whose beaten path skirted the furnace, noticed the next morning that it had tipped into a hole. A knot of Rebel officers, deep in conversation, gathered at the site. As Rose tensely eavesdropped from an upstairs window, he heard the officers repeat the word "rats" several times. The colonel permitted himself to relax slightly; the Confederates evidently blamed the tunneling river rats for the cave-in. But that night, Rose insisted upon going into the east cellar alone, offering up himself for arrest in the event that the Confederates had laid a trap at the tunnel entrance. But he found no evidence that the Rebels had been in the cellar.

He began looking for a place to excavate a new tunnel.

* * *

In the room next to the east cellar, Colonel Abel D. Streight and Captain B. C. G. Reed languished in irons, subsisting on bread and water.

They had fallen into a Confederate trap, and they were lucky to be alive. The Rebels detested Streight for raiding the Deep South, and his written complaints to Union and Confederate officials about the prison conditions had made him the most hated Yankee in captivity. He also sent his friends descriptions of his life in Libby Prison that were often republished in the *National Intelligencer*.

In November 1863 Streight, along with colonels Thomas Rose, William Ely, and William Powell, hatched what became known as "The Great Escape Plot." Streight was the "commander-in-chief," and the others were "brigade commanders." The four referred to the leadership, somewhat confusingly, as the "Council of Five," because the conspirators were organized in "clubs" of five men. (Later, the *Daily Richmond Examiner* would hyperbolically, and mistakenly, pronounce Libby's most famous captive, General Neal Dow, to be "the chosen Moses of the excursion.") Grandiosely conceived as an uprising of all 13,000 Union captives in Richmond, it was to begin in Libby Prison, where, upon a prearranged signal, the Union officers would overpower the guards and seize their arms. They would then free every war prisoner in the city. While one party captured Jefferson Davis and his cabinet, another would ransack the Confederate arsenal. After burning the Tredegar Iron Works and the other James River arms factories, the Yankee escapees would march down the peninsula to the Union lines near Williamsburg.

During October, Libby's black prisoners, utilizing their contacts inside Richmond's servant community, provided the plotters with Rebel troop locations, security arrangements at the arsenal, and other intelligence. Their source was a well-connected "Union lady"—Elizabeth Van Lew or Abby Green, whose servants communicated regularly with Libby's black captives.

As the appointed hour for the uprising neared, to the prisoners' dismay Confederate artillery suddenly appeared outside Libby Prison. The crews aimed the cannon muzzles at the prison walls. At the same time, the Rebel guard was doubled, and General George Pickett's division marched into Richmond.

Someone had betrayed the plot, and its four leaders were forced to abandon it. Recriminations flew. Suspicion soon fell upon Colonel James Sanderson—the former New York hotel keeper and Army of the Potomac commissary. As Libby's commissary chief, Sanderson necessarily worked closely with the Confederates, and consequently many captives believed that he had disclosed the plot to the Rebels.

A few weeks after the scheme's collapse, Streight received an anonymous note with a tantalizing proposition: for the price of two watches and a $100 bribe in greenbacks, Streight and Reed could walk unimpeded through the prison gate. The officers sprang into action. Streight borrowed the cash, Reed obtained the watches, and at a window on the second floor they rendezvoused with Colonel Rose, who brought a coil of rope for them to climb down. Reed went first. When it was Streight's turn, he balked. It is unclear whether he had last-minute qualms or suffered from acrophobia, but Rose finally badgered him into making the descent. Reaching the ground, Streight handed the money and watches to the guard who awaited them, and he and Reed began walking toward the gate. A squad of Rebel soldiers suddenly blocked their path and opened fire on them. The musket balls clattered harmlessly around the Yankees, but they were surrounded and hustled to the dungeon in the middle cellar.

Streight acted as he always did when he believed he had been wronged: he wrote a letter of protest. It was addressed to General Sam Meredith, the assistant Union exchange commissioner, on a page torn from his memo book. The rats and mice were "beyond computation," he told Meredith, and the corn bread, which was the only solid food permitted Reed and him, was "of such a quality that we, as yet, have been unable to eat it." Moreover, "I cannot describe to you the filth, nor the loathing stench with which we are surrounded." Streight slipped the note through a hole in the cellar's low ceiling to a prisoner upstairs. It was copied and given to a Union surgeon who was being paroled.

The raider colonel's friends forced Sanderson, whom they blamed for everything that had happened to Streight, to resign as culinary director. Sanderson and Streight had been feuding since Sanderson disputed Streight's complaints about the prisoners' rations. The feud had spiraled since then, with the warring camps' sniping letters appearing in the *New York Times*.

Streight's friends, in fact, were correct in accusing Sanderson; he *had* divulged the escape scheme to the Rebels, albeit unintentionally. In the days leading up to the planned uprising, Sanderson had been a patient in the Libby hospital. There, he had casually discussed the plot with the hospital "ward master," a civilian war prisoner from Iowa—despite warnings to keep quiet by Colonel Powell of the Council of Five, who was also hospitalized. The ward master repeated Sanderson's remarks to another civilian hospital worker, who relayed the information to the Rebel authorities.

✳ ✳ ✳

TUNNEL NUMBER THREE, LATE JANUARY 1864

Rose and Hamilton had begun a third tunnel. After the castoff brick furnace defeated their last excavation, the digging teams were weary and demoralized, and some of the prisoners contemplated quitting. But Rose and Hamilton would not give up, and the men, as it turned out, required little persuading to make a new beginning, with the small sewer again their objective. "Soon the knives and toy saws were at work again with vigor," reported Lieutenant Frank Moran.

Sensing that their comrades were nearing a breaking point and that they might abandon the project if it lasted much longer, Rose and Hamilton decided to dig day and night to speed the tunnel to completion. Finishing sooner, they concluded, was worth the greater risk of discovery that they courted by working during the daytime. When the exhausted nighttime diggers stole into their rooms at 4 AM, before going to sleep they shook awake a relief crew to take their places in the cellar.

But now the conspirators faced a new challenge: concealing the absence of the men working in the cellar during Erasmus Ross's morning and afternoon appearances. It was not difficult at first. The rooms' connecting doorways had been sealed, but the captives had cut holes in them and crawled from one room to the other, while Ross and his guards had to leave the building after counting in one room, in order to re-enter the adjacent building and count the men in the next room. It was easy to beat Ross into the next room—and there be counted in place of someone at work in the cellar. The captives did not know that Ross was an agent for Richmond's underground, and they considered it "Yankee fun . . . to deceive these Confederates, by means of this small door," according to Captain David Caldwell. But for the tunnelers, it was a game played in deadly earnest. Ross quietly abetted the charade so long as there was no danger of jeopardizing his cover.

Then, at the end of January 1864, five officers escaped from Libby Prison in a single day. After donning civilian clothing, the two groups of escapees had simply walked out of the prison in broad daylight. No guards had challenged them. Four of them were recaptured within a day or two, but Captain John F. Porter of the 14th New York Cavalry safely reached Washington after hiding with Richmond Unionists for nine days.

In response, Major Thomas Turner introduced new security measures throughout the Richmond warehouse prisons. Guards were required to challenge all visitors, to demand identification from strangers, and to detain suspicious characters. They also began routinely bayoneting corpses in the Dead House to discourage captives from shamming. And Ross began conducting a single head count in one room; all Libby prisoners had to be present or face punishment.

No longer able to exploit the crawlways, the tunnelers resorted to a tactic that they called "repeating." After being counted near the head of the line, they would crouch and run behind the ranks of prisoners to the end of the line—to be counted again in the place of their absent comrades. Lieutenant Moran called it "making Ross's book balance." Once, when other officers, thinking they were joining in a prank, raced with the tunnelers to the end of the line to be counted twice, Ross's aggregate exceeded the actual number of prisoners. A captive drily suggested to Ross that Southerners trying to avoid conscription into the Confederate Army were bribing the guards to *get in*.

* * *

The day-and-night excavation work sped the project along; several days after beginning the tunnel, the diggers reached the small sewer. But they discovered that the wood-lined pipe was too small to admit a man. For the moment, they were stymied. As they pondered their options, someone hit upon the idea of dismantling the wooden liner and enlarging the opening, while advancing toward the large sewer.

Removing the liner was difficult, but the captives were galvanized by the notion they were just a day or two from achieving their goal. "It was confidently believed that an entrance to the main sewer would be gained on the night of January 26," wrote Lieutenant Moran. Two diggers were deputized to range ahead and break through the planks lining the main sewer, and to explore it and its outlets.

But the scouts returned from their reconnaissance with discouraging news: the large sewer's lining was made of seasoned oak—rock hard and three inches thick. There would be no easy entry into the sewer. They would have to saw and chip their way in, just as they had cut through the massive supporting timbers while digging the first tunnel.

The tunnelers took up their tools and assaulted the stout oak. Muddy sewage began to ooze into the tunnel, creating an unprecedented stench. One man collapsed to the dirt floor in a faint.

The prisoners' chisels barely dented the tough oak, and their worn-out penknives snapped. Despite the fanners' prodigious efforts, there wasn't enough air inside the tunnel to keep the candles lit.

The overpowering odor sickened most of the men to the point that they could not work and, one by one, the others lost heart.

The men informed Colonel Rose that they could no longer continue. Abandoning their tools, they climbed the rope ladder into the kitchen fireplace, possibly for the last time.

"All the labor expended had been in vain," lamented Captain I. N. Johnston. "The feelings of that little band, who can describe! From hopes almost as bright as reality, they were suddenly plunged into the depths of despair."

For Colonel Rose and Major Hamilton, the failure of the third tunnel was a setback, but not a defeat. They had labored longer and harder than any of the captives—thirty-nine days from the night that Hamilton had loosened the first brick in the kitchen fireplace. The two officers refused to abandon their quixotic undertaking. A war prisoner, they fervently believed, was duty-bound to attempt to escape.

They began making a new plan.

9

General Butler's Raid

"You will see that the prisoners are to be sent away to Georgia. Now, or never, is the time to strike."

–General Benjamin Butler to War Secretary Edwin Stanton

"We are starving by inches."

–Lieutenant Cyrus P. Heffley

FORTRESS MONROE, FEBRUARY 4, 1864
Hat in hand, the road-weary servant boy presented himself to the paunchy, balding, baggy-eyed commander of the Union 18th Army Corps, General Benjamin Butler. Undoubtedly relieved to put behind him the perils that might have cost him his life, the seventeen-year-old handed the general a token that verified his trustworthiness. Then he presented a letter entrusted to him in Richmond, which he had promised to deliver personally to Butler.

The enciphered letter, dated January 30, was addressed to James Ap. Jones, and signed by his niece, Eliza A. Jones. It was Elizabeth Van Lew's first dispatch to Butler since they began corresponding in December. Few of Van Lew's messages would be as consequential as this one. As an aide decoded it, Butler questioned the teenager, and a secretary took notes.

The servant had stayed about a week with Elizabeth Van Lew, who had patiently coached him in memorizing information gathered by a

Unionist source, Charles Palmer. Among other things, Palmer had amassed enemy troop estimates for Richmond and its surrounding area, along with information on the location of the Rebel positions. The servant now gave a compendious recitation of this precious intelligence.

Palmer believed that "Richmond could be taken easier now than at any other time since the war began." Not only was the city lightly defended, but reinforcements would be slow to arrive; it would take at least four days for the 25,000 to 30,000 Rebel soldiers scattered throughout central Virginia to mount an effective response. Butler must act quickly, and in concert with a feint toward Petersburg and a diversion by General George Meade along the Rappahannock River. Van Lew had instructed the boy to warn Butler, however, "not to undervalue Lee's force," and not to attempt any movement against Richmond with fewer than 10,000 cavalry and 30,000 infantry.

If this information did not prod Butler into action, Van Lew's letter did. Once they had been deciphered, the spy's opening words made Butler sit up straighter: "It is intended to remove to Georgia very soon all the Federal prisoners; butchers and bakers to go at once."

This was shocking news. The Confederates planned to move the 13,000 Union war captives in Richmond to the Deep South—beyond the reach of possible rescue by Union forces.

The massive prisoner transfer was imminent. A spy nicknamed "Quaker"—possibly William Rowley, who was not really a Quaker but a Dunker, and a member of Richmond's underground—reported that Confederate kitchen workers and other support personnel had already been selected, and artillery batteries were being placed along the presumed route of march.

In a telegram to War Secretary Edwin Stanton, Butler recommended an immediate attack to free the Richmond captives. "You will see that the prisoners are to be sent away to Georgia. Now, or never, is the time to strike." Butler proposed to "make a dash with 6,000 men, all I have that can possibly be spared. If we win, I will pay the cost; if we fail, it will be at least an attempt to do our duty and rescue our friends. . . ."

With Stanton's blessing, Butler began preparations for a lightning stroke against Richmond.

* * *

Even before General Butler read Elizabeth Van Lew's enciphered letter warning of the Confederates' plan to move the Union prisoners to the Deep South, he had contemplated a raid on Richmond. Four days before her messenger reached him at Fortress Monroe, Butler sent a cipher letter to General Henry Hayes Lockwood: "Can you find me in Baltimore a plan of Richmond? I may want it some time or other, and have not one."

Van Lew's letter, however, was the galvanic shock that pushed Butler from thoughts and words to action. On the day of the letter's arrival, February 4, he ordered General Isaac J. Wistar to prepare to swoop down on Richmond and "relieve our prisoners who must otherwise, it seems to me, of necessity be starved." While Butler told War Secretary Stanton of the Rebel plan to transfer the captives, his instructions to Wistar did not mention this intelligence—possibly as insurance against a potential leak that could jeopardize Butler's Richmond contacts.

Butler ordered Wistar to free the Richmond captives and to "destroy the public buildings, arsenals, Tredegar Iron Works, depots, railroad equipage and commissary stocks of the Rebels." After accomplishing these objectives, he was to kidnap the Confederacy's civilian leadership. The capture of Jefferson Davis and his cabinet, wrote Butler, would deal the Confederacy "a blow . . . from which it will never recover."

The electrifying news of an impending raid on the Rebel capital spread through the Union forces at Fortress Monroe and the chain of camps and forts extending northwest to Williamsburg. The prospect of action thrilled the men of the peninsular army, who had not struck at Richmond since General George McClellan's failed campaign nearly two years earlier.

* * *

The Northern press, too, was urging the Union command to march on Richmond and free the captives. "They must not be permitted to perish," wrote the *New York Herald*. "But what is to be done? That which should have been done long ago. Those prisoners at Richmond must be rescued by force of arms; and this can be done." The *Herald* proposed that President Lincoln raise 50,000 militiamen for sixty days, so that the regular Union Army troops who were guarding Washington could launch an attack into Virginia to liberate the captives. Politicians, educators, and the parents of soldiers got into the act, demanding that Lincoln bring home the war prisoners.

The uproar's apotheosis came in January 1864, when the U.S. Senate, after cataloging the Rebels' sins against the Union captives, recommended mustering one million volunteers for ninety days "to carry food and freedom to every captive held in rebel prisons, and to plant the flag of the United States upon every prison they occupy." The Senate resolution went on to pledge that upon Congress's adjournment, every member under fifty who voted for the resolution would report for military duty in the new units. Whether because of its grandiosity, or its commitment of the senators to military service, the resolution died in committee.

* * *

General Robert E. Lee had recommended moving the Richmond captives to permanent prisons farther south. Richmond was not only unsuited for war captives, their very presence in the capital was "very injurious," Lee told War Secretary James Seddon in a letter on October 28.

Lee understood that the cartel's suspension had closed the Confederate prison system's release valve, and that the 13,000 captives, with more arriving weekly, posed a physical threat to Richmond, while offering a tempting target to Union raiders. The prisoners taxed the capital's limited food supplies, drove up prices, and placed heavy demands on the South's fragile rail system, wrote Lee.

As Lee was recommending the prisoner transfer, General John Winder, the commander of Richmond's prisons, was monitoring a sharp increase in prisoner deaths on Belle Isle. More than a hundred prisoners were now perishing each month. Medical director William A. Carrington blamed the deaths on overcrowding, as well as worsening food and medicine shortages at Belle Isle and Libby Prison. Move the captives to "a more Southern climate," Carrington suggested.

Winder had reached the same conclusion. On November 24, he dispatched his son, William Sidney Winder, to canvass the Deep South for a place to build a spacious new war prison.

* * *

William Sidney Winder's travels yielded three potential sites in Georgia. Fierce local opposition to his top two choices caused him to fall back on his third alternative, 16 acres between Plains and Americus. There, under the supervision of William Sidney Winder's cousin, Captain Richard Winder, workmen in mid-January began building a stockade prison. Labor and material shortages, however, delayed the erection of the barracks, hospital, and bakery. Prisoners began arriving before the project could be completed. The three Winders had planned a prison for 10,000 captives, but insufficient materials and the premature torrent of Union prisoners spoiled their design.

While its official name was Camp Sumter, the new prison would forever be remembered, with sorrow, rage, and horror, by the name of the adjacent town—Andersonville.

* * *

In Richmond, prices for rice, corn, and flour—when they were available—reached new heights. The city's bloated population, the Union blockade, transportation disruptions, and rampant speculation all contributed to the inflated prices. "You take your money to market in the market basket and bring home what you buy in your pocketbook," wrote Mary Chesnut, quoting a popular maxim.

Elizabeth Van Lew set out one late January day in search of corn meal and found none. At City Market, the vendors told her that "the people would come crying to them for meal and they did not know what to do. . . . There is starvation panic upon the people." John B. Jones, the War Department clerk, observing that he could now see the outline of his ribs through his skin, wrote, "We are all good scavengers now, and there is no need of buzzards in the streets. Even the pigeons can scarcely find a grain to eat."

The subject of civilian unrest evidently arose in late January inside the War Department. Jones described a nightmare scenario in his journal: "A riot would be a dangerous occurrence, now: the city battalion would not fire on the people—and if they did, the army might break up, and avenge their slaughtered kin."

On January 19, someone set fire to a woodpile in President Jefferson Davis's basement and nearly burned down the Confederate White House. Detectives suspected that Unionists had an ally in the White House, but they made no arrests. The brazen deed heightened fears of a slave insurrection.

The prospect of another year of war without any hope of peace weighed on the people. War had "made huge gaps in our home circles, had multiplied the vacant chairs around our hearthstones," wrote Sally Putnam.

In their rage with the progress of the war, high prices, and food shortages, Richmond citizens turned to a favorite scapegoat—the Jews. Because they were merchants, they served as convenient targets, although most of them were jewelers or dry goods' merchants, and did not sell food. That did not stop boys from smashing synagogue windows with rocks. Nor did it prevent the magazine *Southern Punch* from proposing a new name for Richmond, "Jew-rue-sell-'em," and referring to Jews as "Richmond Yankees"—patently unfair, as Jewish men served in the Army of Northern Virginia. When many Jews wearied of the abuse and decamped from Richmond, residents sneered that they had "battened and fattened upon speculation" and were now "fleeing from Richmond with the money they have made."

✷ ✷ ✷

LIBBY PRISON

It was another demeaning sightseeing tour. Prison officials were escorting a party of tourists through the living quarters, and pointing out some of the better-known prisoners as if they were zoological specimens. As usual, the captives retorted with mocking cries of "Fresh fish, give him air!"

But these visitors were of a different order, and the cries ebbed as the captives studied them with frank curiosity—and grudging respect. Among a group of Confederate officers, they recognized General A. P. Hill, who had commanded Rebel troops on every major Eastern battlefield, and General John H. Morgan. Weeks earlier, the famous Rebel raider had electrified the South by escaping from the Ohio State Penitentiary.

Morgan and his men were captured while raiding in Ohio—much as Colonel Abel Streight and his men were run to ground in Georgia—and the general and thirty of his officers were sent to the state prison in Columbus. It was rare for a war captive to be imprisoned with common criminals, but Morgan was a special case. Between October 1862 and his capture in July 1863, Morgan's slashing raids had disrupted Union Army operations in Kentucky and Tennessee, destroyed millions of dollars worth of property, and seized nearly two thousand Union soldiers.

Wrongly presuming that Colonel Abel Streight and his "mule brigade" raiders had been sent to Southern penitentiaries, Union officials endeavored to treat Morgan and his men with commensurate harshness. As they did with all new convicts, Ohio officials shaved the Rebels' heads to rid them of lice, and they then mostly ignored them.

The Confederates explored every inch of their cells and discovered that a ventilation shaft ran beneath the floor. With knives pilfered from the dining room, they chipped holes through the cement floor and two feet of masonry into the air chamber, and then dug a tunnel that terminated beyond the prison's outer wall. After laying crude dummies in their beds to fool the guards, Morgan and six companions escaped. The escapees thoughtfully left a note for the warden. It explained that they had commenced work November 4, and had labored three hours a day until the job was finished on November 8. Five of the seven escapees, including Morgan, reached the Southern lines. The "Marion of Kentucky" was warmly welcomed in Richmond.

Standing before the gawking Libby prisoners, the legendary raider searched in the crowd for General Neal Dow who, before Morgan escaped, was to have been exchanged for him. When he found the sexagenarian temperance crusader, Morgan smiled. "General Dow, I am very happy to see you; or, rather, I should say, since you are here, I am very happy to see you looking so well." Dow didn't hesitate. "General Morgan, I congratulate you on your escape; I cannot say that I am glad you did escape, but, since you did, I am pleased to see you here." The ice broken, Morgan told Dow that he was "surprised and sorry" to see the Union officers in such poor condition, and promised to speak to the Confederate authorities. After Morgan left, the favorably impressed prisoners described him as tall, "gentlemanly," and a "generally . . . fine-looking man."

✳ ✳ ✳

Indeed, the plight of Libby Prison's captives reached a nadir in January 1864. "We have never been reduced to so wretched a condition, with regard to provisions," lamented Colonel Federico Cavada. "Empty shelves and empty boxes meet the eye everywhere." "Many are utterly broken down with despair," wrote Benjamin Booth, who wondered whether the Union government "will leave us here to rot and die."

Many prisoners did not survive the winter of 1864. When a morning roll call revealed that a man was missing, Major Thomas Turner stalked around the room looking for him. Spotting a blanket-covered figure on the floor, Turner kicked the prisoner, and ordered him to turn out. When this elicited no response, Turner looked more closely at him, and exclaimed, "My God, I've been kicking a dead man."

Cold, hungry, and lice-ridden, denied relief packages and barred from the market, the Union captives in February struggled to make do. The cold was "as bitter as in Minnesota or Wisconsin," a captive from the Midwest observed. "The sufferings of the prisoners in the upper rooms is indescribable, owing to the want of blankets and clothing," noted Willard Glazier. On the many mornings when there was no firewood, the officers assembled in the kitchen and, in columns of four, stepped out at "a healthy double-quick," often led by General Neal Dow. In desperation, the captives even burned the wooden partition around the privy for warmth and to boil their coffee. The Rebel sentinels suffered, too; when they called out their posts and the hour on these freezing nights, one sometimes heard, "Post number 16, two o'clock, and cold as hell!"

Half of January passed without the prisoners receiving meat. When a Confederate quartermaster ordered meat to be purchased from the city markets for the starving prisoners, War Secretary James Seddon countermanded the order; the meat, he said, must instead go to Rebel troops. Libby's adjutant, Lieutenant John Latouche, bluntly told the captives that he wished he were rid of them, because he couldn't feed them. Then, to the captives' amazement, the guards on post near the windows began begging for bread. "I am afraid the Confederacy will rot down over our heads," wrote Captain Robert T. Cornwell.

Waging their own solitary battles with despair and hunger, the Rebels became even more resentful of the captives. "The rebel guard[s] say that the 'Yankees are dying right smart now at the hospitals,'" Sneden reported. One night in Libby, the captives' singing and marching at double-quick time in the kitchen in order to stay warm brought a squad of armed guards. A highly irritated Major Thomas Turner up-braided them for "howling . . . your villainous Yankee songs." As punishment, they were made to stand in ranks until midnight.

When they lay down to sleep at night, the captives listened to the Rebel soldiers smashing open their undelivered boxes. "I hope someone will set fire to the building in which the boxes are . . . 15 or 20 ton have been laying there for over a month. We are starving by inches," Lieutenant Cyrus P. Heffley wrote bitterly in his diary. General Benjamin Butler and Colonel Robert Ould discussed resuming the box deliveries on both sides. But they reached an impasse when Ould demanded that "copperheads"—Southern sympathizers in the North—be permitted to bring food to the imprisoned Rebels.

On February 1, Turner curtailed one of the captives' last consolations—their correspondence with loved ones. "Only one letter to each officer will be allowed to be sent to the *so called* United States each week," he announced. Moreover, each letter was limited to personal subjects of no more than six lines. The Rebel censors were overwhelmed by the blizzard of mail produced by the 13,000 Richmond captives, many of them prolific letter-writers—each letter requiring inspection for codes, hidden messages, and evidence of "invisible ink" before being sent on. While the purpose of Turner's new rule ostensibly was to ease the censors' workload and facilitate the letters' delivery, the captives later learned that many of their weekly six-line letters simply vanished.

* * *

Colonel Cavada, a close observer of humanity and a philosopher, described captivity as "a flail which threshes the chaff out of human pride." During the starvation months of January and February 1864, it was easy to die in Libby Prison, and not everyone possessed the will to stay alive. Army engineer Warren Lee Goss witnessed the death of a young, fair-haired prisoner, covered with flies and lying in "liquid filth." Goss watched the young man raise a filthy piece of bread to his mouth—and expire. A one-legged prisoner dragged himself to the body, unclasped the fingers still holding the bread, and ate it like "a famished wolf." At times, wrote Lieutenant George Putnam, all that stood between the starving captives and death was "this will power, the decision to live if possible, the unwillingness to give up, beaten by the Confederacy or by circumstances."

But even for the strong-willed, it was difficult to remain indomitable. A. C. Roach, an aide to Colonel Abel Streight, confessed that on many nights in Libby, he lay down on the floor and "devoutly prayed that I might remain in the unconsciousness of sleep until the day of deliverance from my wretched condition." Chaplain Charles McCabe overheard a dejected Irishman singing to himself, "Backward, turn backward, O Time, in your flight / Make me a child again for just to-night," and another Irish prisoner's riposte, "Yis, and a girrul child at that!"

Those who succumbed to illness or disease, starvation or despair were stacked like flour bags in Libby's west cellar, where they were prey to every kind of carrion-eater. When there were enough corpses to make up a "load," black prisoners stuffed the bloated, mutilated bodies into coffins for the final trip to Oakwood Cemetery in a mule-drawn "dead cart." A Confederate soldier would call to the driver, "A load of dead Yankees! Drive up your mule!" By surreptitiously marking the coffins, the prisoners confirmed their suspicion that the coffins were being reused. But by early 1864, so many prisoners were dying that the Confederates dispensed with even that thin pretense; when the dead cart arrived—every day now—up to a dozen bodies were slung into it, like butcher carcasses. A prisoner mordantly observed, "So we die like dogs, and are buried like dogs!"

Yet, the deprivations and suffering of some Southerners approached those of the prisoners. Every day, barefooted children assembled beneath Libby's windows, clutching their thin outer garments close to their bony frames. "We throw to them spare fragments of corn bread, and occasionally a macerated ham bone, which they scramble for greedily, to carry home with them," wrote Colonel Cavada. Sometimes adults joined the child beggars. "The poverty-stricken begged from the impoverished," observed Captain Bernhard Domschke. Cavada reported that an enterprising cow also began showing up in the hope of receiving some of the thrown scraps. On the occasions when she was ignored, she shook her head so the bell around her neck clanged loudly, while "gazing wistfully up at the barred windows."

* * *

WILLIAMSBURG, VIRGINIA, FEBRUARY 6, 1864

After nightfall, General Wistar led six thousand mounted men and two light batteries out of Williamsburg, bound for Richmond. Dust churned along the Williamsburg Road between the Pamunkey and Chickahominy rivers as the raiders rode through the moonlit Virginia low country. Each man carried six days' rations and seventy rounds of ammunition. Wistar planned to cross the Chickahominy twelve miles below Richmond at Bottom's Bridge, which he hoped would be unguarded. After crossing the river, the raiding party would proceed through Seven Pines and enter the capital's southeast suburbs. As the hours went by without any sign of enemy troops, the raiders began to believe they had caught the Confederates napping.

At 2:30 AM on February 7, they reached Bottom's Bridge—and to their horror discovered that it had been destroyed. Moreover, Rebel troops were dug into defensive positions on the opposite riverbank, and trains were busily bringing reinforcements from Richmond. By daybreak, several Confederate infantry regiments, a cavalry unit, and four artillery batteries guarded all of the nearby river fords.

A Union detachment tried to force a crossing, and was repulsed with nine casualties. Frustrated and puzzled by the Rebels' obvious foreknowledge of the raid, Wistar turned his men around and returned to Fortress Monroe.

In Richmond, the fighting at Bottom's Bridge sounded like a distant thunderstorm, and it sent the city into pandemonium. Church bells rang madly, and people mobbed the streets. At Libby Prison, the captives laughed at the "ridiculousness" of the scene, and jeered as two mountain howitzers went by, pulled by bony, swaybacked mules. Small boys and old men ran to join veterans, raw militia, and frightened citizens in defense of the city. The jubilant prisoners believed their day of deliverance was at hand, wrote Adjutant S. H. M. Byers of the 5th Iowa. Evidently, the Rebels thought so, too, because howitzers suddenly appeared on the nearby streets, aimed at the warehouse prisons and the James River bridges.

* * *

A few days later, Butler read a *Daily Richmond Examiner* article that said a "Yankee deserter" had tipped the Rebel command to the raid. Butler ordered Wistar to find out who he was. Wistar's investigators learned that a guard at Fort Magruder near Williamsburg, Private Tom Abraham of the 139th New York, had accepted a bribe to aid the escape of Private William Boyle of the New York Mounted Rifles, who was under a death sentence for murdering an officer. In their whispered conversations before Boyle slipped away, the condemned man elicited from Abraham all that he knew about the impending raid. Eluding the Union pickets at Williamsburg, Boyle entered Rebel-controlled territory, and then had no trouble reaching Richmond—where he reported to the Confederate command everything that he had learned from Abraham.

Boyle was still alive in early February only because President Lincoln two months earlier had suspended all military executions until further notice. Lincoln was never comfortable with Union troops being put to death, and he sometimes delayed sentences until he determined for himself whether the punishment fit the crime, often granting clemency. The moratorium gave Boyle time to escape and divulge Butler's plans to the enemy.

The discovery of Private Boyle's escape and subsequent treachery culminated in Abraham confessing to everything. "A prompt example [must] be made," Butler wrote. "A higher military crime cannot be conceived." The general might have asked why a lowly private on guard duty knew enough about a supposedly secret, high-risk operation to cause its failure. But Butler did not question the mission's porous security or demand new safeguards to prevent a recurrence.

He did attempt to obtain custody of Private Boyle. "As this man is a murderer duly convicted, it is believed the Confederate authorities will not desire to retain him, as a murderer is defined to be an enemy of all mankind," Butler wrote to Colonel Ould. He offered to exchange a Rebel prisoner for Boyle. The offer was declined, and Boyle was not returned. The Rebels knew that giving him up would spoil their chances of receiving intelligence windfalls in the future.

Exactly a month after Wistar's raiders were turned back at Bottom's Bridge, Abraham was taken to a field near Yorktown. In a hollow square formed by two lines of troops, artillery batteries, and the York River, a firing squad poured a volley of musket fire into Abraham as he sat blindfolded on his coffin.

A representation of Libby Prison life before prisoner exchanges were suspended and crowding and food shortages became endemic. *(Virginia Historical Society)*

Colonel Abel D. Streight. Captured while leading a Union raid in northern Alabama and Georgia, the Indianapolis publisher was loathed by Confederates more than any other Union war prisoner. *(Library of Congress)*

Colonel Thomas E. Rose. Escape was uppermost in the Pennsylvania schoolteacher's mind from the moment he arrived in Libby Prison. *(Library of Congress)*

Major A. G. Hamilton. The Kentucky homebuilder discovered an ingenious way to reach Libby Prison's east cellar from the prison kitchen. *(Library of Congress)*

General Benjamin Butler. "Beast Butler" was hated in the South for his conduct in New Orleans, but the Confederacy was forced to negotiate with him about prisoner exchanges. *(Library of Congress)*

City Point, Virginia. Located on the James River near Petersburg, this was where thousands of Union and Confederate war prisoners were exchanged. *(Library of Congress)*

Elizabeth Van Lew. A member of Richmond's social elite, she was the "guiding spirit" of the Union underground, aiding escaped war prisoners and supplying the Union command with valuable intelligence. *(Virginia Historical Society)*

The Van Lew mansion on Church Hill in Richmond. *(Virginia Historical Society)*

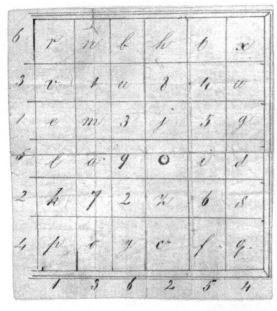

Elizabeth Van Lew's cipher key, a matrix of thirty-six seemingly random letters and numbers. She consulted it when composing and deciphering correspondence with the Union command. *(Elizabeth Van Lew papers, Manuscripts and Archives Division, New York Public Library, Astor, Lenox and Tilden Foundations)*

In this February 1864 notice, Major Thomas Turner, the Libby Prison commandant, restricted the Union captives to just one weekly letter of no more than six lines. *(George W. Grant Papers, Duke University Special Collections Library)*

Digging the Libby Prison tunnel. The man on the left is turning to empty dirt from a cuspidor; the prisoner on the right is fanning air into the tunnel. (*Century Illustrated Monthly Magazine, March 1888*)

This illustration shows Colonel Thomas Rose emerging from the just-completed tunnel. Note the chisel in his right hand. (*Century Illustrated Monthly Magazine, March 1888*)

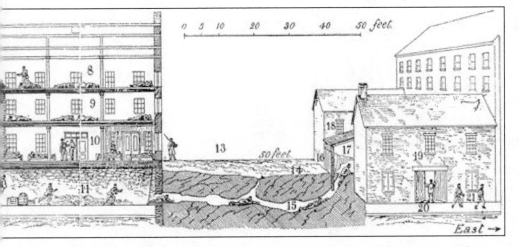

An imagined cross-section sketch of the east side of Libby Prison and the escape tunnel. (*Century Illustrated Monthly Magazine, March 1888*)

A dramatic re-creation of the capture of Libby escapees by Confederate cavalrymen. (*Library of Congress, from Harper's Weekly, 1864*)

In 1888, the Libby warehouse was dismantled, and its 600,000 pieces were loaded into 132 boxcars and transported to Chicago. It was rebuilt as the Libby Prison War Museum, pictured in this postcard. *(Library of Congress)*

Elizabeth Van Lew's headstone in Shockoe Cemetery in Richmond, Virginia. The bronze tablet inscription reads: "She risked everything that is dear to man—friends / fortune, comfort, health, life itself, all for the / one absorbing desire of her heart—that slavery / might be abolished and the Union preserved." *(Pat Wheelan)*

10

The Ordeal of Tunnel Four

"From this time forward he [Rose] never once turned over the chisel to a relief. All day long he worked with the tireless patience of a beaver."

—Lieutenant Frank Moran

LIBBY PRISON, LATE JANUARY 1864

Left to carry on alone, Rose and Hamilton decided that tunneling toward the Kanawha Canal, and downward through wet clay, was impracticable. Yet they still believed that a tunnel offered their best chance of escaping, and so they decided to strike out from the east cellar toward a new objective.

If they began in the northeast corner, they would not have to dig below the cellar floor before tunneling out, as was the case in the southeast corner, where the ground surface sloped downhill toward the canal. The northeast corner lay below the ground surface, yet above the water table, so it would not flood and the clay would be firm and dry.

But Rose and Hamilton foresaw new complications. The sewer's advantages were its proximity to the cellar and its invisibility; the diggers might have reached the sewer and the canal bank without being seen. By keeping to higher ground and tunneling eastward, the diggers would have to pass beneath a broad dirt yard patrolled by Rebel guards. Only on the yard's far side, more than 50 feet distant, was there hope of concealment in the form of a tall fence and a tobacco shed. But the captives would still be on Libby property, beside the building where the undistributed boxes were stored. It would not be easy to leave the premises with the Confederate sentries patrolling the nearby streets all day and night.

Near the storage building was an arched entryway onto Canal Street, which adjoined Libby's southern boundary and ran beside the canal. Wagons delivered goods to the prison through the gate by day; at night, the gate was closed and illuminated by gas street lamps. A guard stolidly paced Canal Street, stopping and reversing direction before reaching the wagon entrance. Before performing his half-pirouette, the guard had a clear view of the archway.

These potential obstacles notwithstanding, Rose and Hamilton decided to begin a new tunnel that would take them to the fence and shed. If they advanced their project to the point that their final concern was entering Canal Street unobserved, they would find a way.

* * *

When the Confederate guard announced lights out at 9 PM, Rose and Hamilton silently rose from their sleeping places and descended to the kitchen, where they again opened the fireplace chute. In the east cellar's northeast corner, they probed the wall for a likely place to begin digging. After two false starts, they hit densely packed soil and commenced their fourth tunnel.

Rose, Libby's star digger, bore into the wall 6 inches above the cellar floor and in the shadow of one of the fenders that supported the building. If a Rebel guard entered the cellar, from the doorway he would not be able to readily see the workings, and the tunnelers would have time to hide.

For the moment, there were no other diggers. Rose toiled alone in his two-feet-diameter burrow while Hamilton fanned air. He tersely noted, "Hamilton and myself continued our dreary work alone, as before." At first, the ground yielded easily to Rose's chisel and bare hands. But the men soon encountered the same difficulties as before: the candle would not stay lit, and Hamilton could not fan air to Rose quickly enough, while also emptying and hiding the dirt from the cuspidor and keeping a sharp lookout for Rebels. There was no getting around it: it was unlikely that they could complete a 50-foot tunnel without help.

Rose invited the former tunnelers to a meeting. After describing the excavation that he and Hamilton had begun, he persuaded most of the men to rejoin the effort. They resumed their day and night shifts and their former practice of "repeating" during Erasmus Ross's head counts to mask the absences of the daytime diggers.

One day, a loud crash resonated through the prison; the daytime digging team had breached a foundation wall far more noisily than intended. Prison officials dispatched guards to find out the cause of the racket. Dread filled the tunnelers in the upstairs rooms. But they were lucky. Workers retrofitting the prison's windows with iron bars had made the racket, the guards reported. When the news reached the tunnelers, "our boys thumped away with a will without exciting the least suspicion," wrote Captain I. N. Johnston.

* * *

There was a sense among the diggers that this tunnel was the one. Perhaps it was Rose's encouragement as they advanced a half inch with each chisel stroke. Or it might have been because the conspirators' luck was holding during the twice-daily prisoner tallies—the captives never suspecting that Ross was studiedly ignoring strange behavior that would have aroused the suspicions of a committed Confederate.

The tunnel descended at a shallow gradient before leveling out, conforming to the contours of the ground surface eight feet overhead. At its portal, the passageway was 24 inches by 18 inches, but at one point its width narrowed to just 16 inches. The diggers wriggled in and out of it, working face down, flat on their bellies. Since Rose had nearly lost his life, they worked with a rope knotted around one leg so they could be dragged out if it became necessary.

The new tunnel was now longer than any of the earlier tunnels, and the diggers used up its oxygen even more quickly, periodically returning to the cellar to recover. "It was impossible to breathe the air of the tunnel for many minutes together," wrote Johnston. Panting from their exertions and the lack of air, the diggers fought the sensation of suffocation that hovered on the edge of their consciousness. "We often pulled out our comrades, suffocated and exhausted, nearer dead than alive," reported Captain W. S. B. Randall.

The worsening conditions inside Libby Prison spurred on Colonel Rose and his tunnelers. "The hard fare and confinement of our prison, the monotony of which had become unendurable, and the possibility of escape at last roused us up to exertions almost superhuman," wrote Johnston. "There are no impossibilities to men with liberty as the result of their labors." The conspirators pushed the fourth tunnel steadily forward, even as new dangers loomed.

As part of Major Turner's security crackdown, Libby guards now inspected every corner of the prison. The tunnelers were no longer safe from discovery in the east cellar. At any moment during the daytime, a Confederate sergeant and several soldiers might enter the cellar through the south door while the tunnelers toiled in the room's shadowy north side. During these terrifying visitations, the digging team scattered to hiding places during the precious seconds that it took for the Rebels' eyes to adjust to the darkness. As the digger, his head protruding from the tunnel, tensely watched the intruders, the others crouched behind the building pillars or hid under the straw. The guards seldom ventured into the cellar's north end, and the stench, the squealing rats, and the sepulchral darkness hastened their departure. Nonetheless, off-duty tunnelers began standing watches in the upstairs rooms to alert the diggers to approaching guard details.

* * *

EARLY FEBRUARY 1864

Something was different about this morning's prisoner tally. As Colonel Rose's tunnelers prepared to cover for the two men who were at that moment at work in the new tunnel in the east cellar, Erasmus Ross strode into the Lower Chickamauga Room with twice the usual number of guards, and with several Confederate officers, including adjutant Lieutenant Latouche. The nervous tunnelers got into formation as the guards and officers fanned out around the room.

Ross began conducting a head count. The prisoners' prankish "repeating," mimicking the tunnelers' deadly serious efforts to conceal their comrades' absences, may have given the prison clerk an excuse to ignore the subterfuge, but now his superiors were watching. The show of force was designed to deter further shenanigans.

When two tunnelers tried to repeat, the other captives, tired of the "game" and intimidated by the Rebels, refused to let them into the formation. When Ross finished counting, he was missing two men. The prison register was fetched from Ross's office, and a roll call was conducted. When a prisoner's name was read, he passed through a doorway between files of guards.

The roll call confirmed the head count's discrepancy, and identified the missing prisoners as Captain I. N. Johnston and Major B. B. McDonald, both of whom were at that moment in the east cellar. The Rebels began an investigation.

After lights out that night, Rose and the relief tunnel team entered the cellar through the fireplace and delivered the bad news to Johnston and McDonald. They had two options, neither appealing: return to their room, and at the next morning's head count attempt to explain their absences; or remain in the cellar until the tunnel was completed and they could escape. McDonald was confident that he could talk his way out of his predicament and elected to return to the room and take his chances. Johnston, however, chose to remain in the cellar.

* * *

McDonald's appearance at roll call the next morning surprised and perplexed the Rebels, who had presumed that both men had escaped. When informed of McDonald's reappearance, Major Turner sent for him.

With a great show of remorse, McDonald told the commandant that he wished to make a clean breast of what had happened. He had gone downstairs for sick call but then, before going to see the doctor, he had decided to visit a friend in the west room on the third floor. As a result, he had missed the head count and roll call in the Lower Chickamauga Room. It being too late to report to the doctor, McDonald decided to remain where he was until the next morning.

Turner didn't believe him. Ross had searched every room without finding him. How was that possible? With a show of reluctance, McDonald replied that he had been afraid to be found out of his quarters, and so when Ross came looking for him in the upper west room, he had climbed onto the room's joists and lain there on a large plank where he could not be seen from below.

To McDonald's amazement and relief, his explanation satisfied Turner. The commandant sanctimoniously pronounced "an open confession [to be] good for the soul," and permitted McDonald to return to his quarters—without punishment.

* * *

McDonald had risked severe punishment rather than be caged in the east cellar, but Johnston, in choosing the less risky course, had sentenced himself to a long train of days in purgatory. Johnston's fellow tunnelers easily convinced the Rebels that he had escaped. Lieutenant J. C. Fislar concocted a story about how Johnston had escaped several days earlier with the help of two cousins who were Rebel guards. Fislar said the guards had supplied Johnston with a Confederate uniform, and Johnston had made a wooden sword, tin scabbard, and a belt of oilcloth so that he might pass as a Confederate officer. Other prisoners solemnly corroborated Fislar's account, speculating that by now Johnston had probably reached the Union lines. The Confederates bought the story and suspended their search.

Johnston found his new home almost unbearable. A farmer's son who had fought with the 6th Kentucky at Shiloh and Stones River before being captured at Chickamauga, Johnston was used to an active, outdoor life. Being penned up in Libby Prison had been hardship enough, but he now dwelled in a state of perpetual gloom that was relieved during the daytime by only the thin, chalky light filtering through the south door's iron bars. And he had plenty of time to contemplate and perhaps repent his decision, for Rose had suspended all daytime excavation, deeming it too risky because of the new unpredictability of the Confederates' twice-daily head counts.

The tunnelers brought Johnston food when they descended to the cellar to resume work at night. Sometimes they helped him slip upstairs after lights out to sleep for a few hours among the prisoners in the comparatively pure air of the rooms. This mitigated his misery, but Johnston still spent most of his waking hours "among the squealing rats—enduring the sickening air, the deathly chill, the horrible, interminable darkness."

Lacking anything else to do, Johnston worked fitfully in the tunnel during the daytime, careful not to stray from the cellar's north end, for the Rebel guards' surprise inspections worried him. "I had plenty of company—little of it, however, agreeable, as it consisted of rebels, rats, and other vermin." He became accustomed to the rats, taking notice of them only when they were noisier than usual, which suggested that his "two-legged enemies were near." Beneath a pile of straw, Johnston dug a pit where he concealed the tunneling tools—and himself, too, when the guards came.

* * *

One day, when Johnston was alone in the cellar, the door was suddenly flung open. He dove into his hiding place and covered himself with straw just as several Rebel guards strode purposefully across the cellar, stopping a few yards away from the tunnel entrance. As Johnston anxiously waited for them to leave, the straw dust began to tickle his nose. A sneeze was building behind his eyes and, it seemed, filling his entire head. He tried mightily to hold it back, but it was hopeless. When he sneezed it made such a loud explosion inside his head that he was certain everyone in the cellar had also heard it.

Scarcely daring to breathe, he listened with preternatural intensity, expecting to hear loud exclamations, and at any moment to be dragged from the hole and put in irons. But the Confederates continued to converse in low voices. They were in no hurry to leave, either, and, in fact, seemed to be looking around with more interest than usual. Several of them even casually kicked at the straw that concealed him. As he peered from his hiding place through the latticework of straw, Johnston incredulously watched the Rebels repeatedly pass the tunnel entrance without noticing it. They finally left, without having discovered either Johnston or the tunnel.

After listening to Johnston's account that night, Rose concluded that the Confederates must have learned that the cellar was a staging area for a planned escape, but that they did not know about the tunnel. With the Rebels' suspicions aroused, Rose knew that they had to complete the tunnel quickly.

As the tunnelers inched beneath the surface of the broad yard on Libby's east side, their need to know the distance to the safety of the fence grew more urgent. Captain John Gallagher of the 2nd Ohio, a cheerful, charming Irishman, volunteered to find out. Gallagher somehow persuaded the guards to permit him to cross the yard and hunt among the undelivered boxes from the North for one that had supposedly been sent to him. The Confederates normally denied such requests, but Gallagher had cultivated the guards' goodwill. Trailed by a sentry, Gallagher left the prison for the old tobacco shed, walking at a strangely precise gait—as close to 3 feet per step as he could manage. Of course, there was no box for him in the shed, but Gallagher had gotten what he wanted. He estimated the distance between the Libby warehouse and the tobacco shed to be 52 to 53 feet.

✳ ✳ ✳

SUNDAY, FEBRUARY 7, 1864, 1:00 AM

Advancing 5 feet per night on average, the diggers had now reached a place where they might surface safely—or so believed Colonel Rose and Major Hamilton. Their measurements showed that they had tunneled 53 feet. Colonel Rose instructed Captain W. S. B. Randall of the 2nd Ohio, who on this night had charge of the chisel, to break for the surface. Randall wriggled into the tunnel, and Rose, Hamilton, and the others waited anxiously for him to return with a report from outside Libby's walls.

Randall was digging upward toward the surface, panting for air as dirt fell into his face, when blessedly clean air suddenly poured into the tunnel. But his relief congealed into horror when he saw that while he had emerged beside the fence, he was on the wrong side of it, in plain sight of the prison and the sentries. Worse, a guard who happened to be standing nearby when Randall broke through the surface had heard him dislodge a large stone and was now coming over to investigate.

The guard stopped above the hole that Randall had made and leaned on the fence, peering into the blackness on the other side—the side where Randall wished to be, but was not. For a half minute, the guard stood there. Randall lay frozen, not daring to breathe, his heart pounding, drops of cold sweat standing out on his forehead. "The events of my life seemed to flit before me." And then the guard walked away without having seen Randall's ashen face staring up at him. Randall wormed backward through the tunnel. After the others pulled him into the cellar, he exclaimed, "All is lost!"

Rose wriggled through the tunnel to personally survey the exit. When he returned to the cellar, the colonel reported that the miscalculation, while dangerous, was not necessarily fatal. The guard had not seen the opening, and there was enough time before daybreak to repair the mistake. Moreover, Rose said, the hole had been made where the ground surface sloped downward, and it could not be readily seen from most of the yard. But the hole must be concealed, Rose said, and the tunnel must be redirected to the tobacco shed 5 to 10 feet away. It was Randall who, after covering an old pair of pants with dirt so that they blended with the ground, crawled back into the tunnel and plugged the hole. It had been a close call.

It was a profound disappointment to Captain David Caldwell, who was poised to leave Libby that night, having packed a haversack with a map and food that he had carefully hoarded. He reproached himself for having selfishly permitted the prospect of escape to distract him from his duties as Libby's temporary chaplain. After having balked at a request to speak at a funeral service that day, he now delivered the sermon, while wondering "whether the whole plot had not better be abandoned."

* * *

SUNDAY, FEBRUARY 7, DAYTIME

After a few hours of rest, Rose and Major B. B. McDonald returned before daylight to the cellar, confident that the guards would not be inspecting the area on a Sunday. Before stopping work the previous night, Rose had reoriented the tunnel so that it pointed once more toward the tobacco shed. A day or two more of hard work would finish the job, he believed. Using his hands and the chisel to claw his way blindly forward like a mole, Rose dug with great urgency, trying to make every hour count. He was convinced that it was only a matter of time before the Rebels discovered the tunnel. They had to finish it now.

As Rose dug, McDonald fanned the tunnel entrance, and Captain Johnston maintained a vigil for intruders. From the upstairs windows, the other tunnelers scanned the grounds for any evidence of a surprise inspection by the guards. The "forlorn hope"—American soldiers' nickname for such last-ditch endeavors—had begun.

"From this time forward he [Rose] never once turned over the chisel to a relief. All day long he worked with the tireless patience of a beaver," Lieutenant Frank Moran later wrote.

Rose dug far into the night before withdrawing in exhaustion from the tunnel. Just twelve hours more and the tunnel would be finished, he gasped to McDonald and Johnston. McDonald, who was ill from the reeking cellar and the daylong effort, accompanied Rose as he shakily climbed the rope ladder to the kitchen. They slept until the light of the new day awakened them.

*   *   *

Monday, February 8

It was the seventeenth day since tunnel number four's groundbreaking. From early morning, when they reopened the fireplace and descended the rope ladder to the east cellar, Rose, McDonald, and Johnston worked steadily all day. The cellar's silence was occasionally broken by the scraping of the cuspidor being dragged through the tunnel, and by intermittent rodent squeals, caused by one of their fierce quarrels or a prisoner's boot.

After daytime's gray half-light faded and then vanished, Rose dug on as though possessed. Midnight found him slick with sweat, struggling to control a tremor that shot through his body like a fork of lightning. He was on the verge of collapse. And then he could see that he was passing beneath a fencepost. This was a huge milestone. It meant that the tobacco shed lay just a few feet away. It would not be long now. Panting for breath, Rose began to angle upward, toward the surface.

When his cramped limbs trembled with fatigue, Rose flipped onto his back and kept digging. The hour was at hand. He would break free or die trying. The air fanned into the tunnel by McDonald no longer reached him. Rose was slowly suffocating in the dead air.

He pounded on the roof of his narrow grave. The crust abruptly gave way, and dirt spilled onto his face, choking him. And then he was enveloped in blissfully cool air. The sensation was so wholly alien that he just lay there a moment, staring at the black sky and the stars—so soothing to his burning eyes.

Rose was jarred from his reverie by a sentinel's cry: "Half-past one, and all's well!" Crawling out of the hole, he saw that he was under the tobacco shed, at the place where the high fence shielded him from the prison and the sentinels. His excitement grew. It was perfect.

Keeping to the deep shadows cast by the moon, Rose silently made his way to the latched gate barring the arched wagonway. This was the place that he and Hamilton had weeks earlier determined would be their exit to the street. The massive gate was fastened with an iron bar; exhausted, Rose struggled to lift it and open the gate without attracting the sentries' notice. Peering cautiously around the corner of the gate, he saw that the guard was marching away from him. He stepped into Canal Street.

At this moment of supreme triumph after six weeks of frustration, exhaustion, terror, and near death experiences, Rose might have briefly contemplated making a run for it. But the disciplined colonel reined in his impulses; there were his comrades to consider, too. He permitted himself a single circumnavigation of Libby Prison, carefully avoiding the patrols around nearby Pemberton Prison.

Completing his thrilling stroll, Rose slipped through the wagonway and closed and secured the gate. He crawled under the tobacco shed and lowered himself into the tunnel, painstakingly concealing the exit with an empty hogshead that he had found nearby. Colonel Rose returned to Rat Hell.

McDonald and Johnston were thunderstruck by the news. Rose sent word to Hamilton to awaken the other tunnelers for a meeting in the kitchen.

* * *

At 3 AM on February 9, Rose announced to the men who had labored with him that the tunnel was completed. "They wrung his hand again and again, and danced about with childish joy," wrote Lieutenant Moran.

No one blamed Rose and Hamilton for wanting to leave immediately. Fifty-three days had gone by since they had removed the first bricks from the kitchen fireplace to begin carving the passage to the east cellar. But with daybreak a few hours away, their comrades argued, there was too little time either to escape Richmond or to find a hiding place in the city. If they waited until the next night, they would have ten hours of darkness to cloak their movements. In the meantime, they could rest, and collect the cash, food, and extra clothing that they would need to endure a cross-country journey in wintertime.

Rose and Hamilton saw the sense in waiting. Reluctantly, they agreed to delay their escape until the coming night.

11

Escape

"I felt the soft ground under my feet, and looked over and about me, as if to assure myself that it was not all a dream. I never felt a greater determination to accomplish a purpose in my life."

—Lieutenant James Wells

LIBBY PRISON, TUESDAY, FEBRUARY 9, 1864

Never did a day crawl by as slowly as this one. Colonel Rose, Major Hamilton, and their comrades tried to distract themselves by fussily readying clothing and provisions for the coming night's breakout. "Time here is a burden, a tormentor, a bore," Captain Robert T. Cornwell jotted in his diary on this day. As the hour of their escape neared, Rose announced that each tunneler could bring one friend. The now thirty conspirators spent the day quietly tucking bread, dried meat, cash, matches, and civilian clothing into haversacks for their hegira through the wintry Virginia coastal plain. A few farsighted men had been conditioning their bodies for the ordeal; Lieutenant James Wells and his friends had walked in laps around their room, increasing their distances until they covered 22 miles in one day, by their estimation. But Wells's footwear was in such disgraceful condition that he was nearly barefoot, and he spent part of the day trying to persuade a Lieutenant Mead to swap his boots for some of Wells's "choicest things." Mead was uninterested in making a trade.

The captives knew that once they left Libby Prison and entered Canal Street, they would face a galaxy of fresh dangers: hostile civilians, enemy soldiers, partisans who would gladly shoot them, and trackers with blood-hounds. At the same time, they would have to endure unremitting exposure to cold and frost, and possibly snow and freezing rain, while wading through swamps and sleeping in the open. But most of them regarded the looming hardships as no more than the elements of a grand adventure that would be crowned by their liberty. Their chief concern during the interminable day was curbing their excitement. They could not allow their emotions to betray them, with liberation so near at hand.

Outside Libby's walls, it was a bright, chilly day, warming to just 43 degrees. Richmond had returned to its usual state of wartime agitation following the brief panic caused by General Wistar's failed raid two days earlier.

In Washington, a photographer took a picture of President Lincoln as he was about to attend a large White House levee. The photograph would grace the $5 bill sixty-four years later.

* * *

No prisoner looked forward to nightfall more eagerly than Captain I. N. Johnston, for whom escape meant not only liberation from Libby Prison, but from the east cellar, his place of exile. "Never was my anxiety so great as for the setting of that day's sun," he wrote, "and more than once during its long, dreary hours I feared that the cup of happiness, now so near our lips, would be rudely dashed away."

Johnston's anxiety was justified; the Confederates, evidently told by informers that the prisoners were planning something, had been inspecting the east cellar with disconcerting frequency. Each visit was a terrifying experience for Johnston, who had to dive into his hiding hole beneath the straw pile and pray the guards would not find him or the nearby tunnel entrance. Once, a guard brought a dog into the cellar, and sicced it on the rats. "In his search after them he passed over and round me. . . . I was too large game for him."

* * *

Colonel James Sanderson, the ex–commissary officer and pariah, was a popular man at the moment because of his willingness to lend cash to the conspirators. Although his indiscretion had spoiled the Great Escape Plot, Sanderson was nonetheless considered trustworthy enough to ask for a loan. One tunneler even invited Sanderson to be his escape companion, but the colonel declined. Sanderson gave money to Major B. B. McDonald—who would travel in the company of Colonel Abel Streight, Sanderson's mortal enemy—Lieutenant R. W. Randolph, and others. He also donated a hatchet and, to the conspirators' delight, a bottle of brandy. Standing at a window with McDonald, Sanderson pointed to a house on a hill where, he said, McDonald would find a Unionist woman willing to aid him. Undoubtedly, Sanderson was referring to Elizabeth Van Lew's Church Hill home, which could be seen from Libby Prison. In fact, McDonald had already been in contact with Van Lew.

They might have supported the escape party remuneratively or in spirit, but the family men balked at risking themselves and, by extension, their loved ones' futures. Older men like Sanderson feared they would slow down the others. And there were other captives, wrote Captain Charles Earle, who held the opinion that "they were captured in the line of duty by no fault of theirs, and if the Government needed them it could effect an exchange." Others, incapacitated by long months of inactivity, short rations, and illness, doubted their ability to survive as fugitives in the countryside. Colonel Frederick Bartleson had lost an arm in battle and believed that it would be useless for him to attempt to navigate the fireplace and tunnel, "if not decidedly detrimental to others." General Neal Dow excused himself on the grounds of age and debility. "My naturally strong constitution had been broken down," he wrote. In his case, this may have been an excuse; Dow knew he would soon be exchanged.

* * *

While they waited for nightfall, the tunnelers quietly discussed the escape. They unanimously decided that they were entitled to leave before everyone else, because they had fought off the rats, and had endured the tunnel cave-ins and close calls, and the stench and exhaustion. They designated Colonel Harrison C. Hobart of the 21st Wisconsin as the fireplace gatekeeper. After the fifteen tunnelers descended to the cellar, he would retract the rope ladder and seal the fireplace. An hour later, Hobart would then reopen the fireplace to admit the tunnelers' fifteen chosen friends. The thirty men would be required to whisper a password before Hobart would allow them to pass into the cellar, and before a second gatekeeper in the cellar would let them into the tunnel. Thereafter, at regular intervals Hobart would open the fireplace to other prisoners, priority being given to the highest-ranking officers. Each flight would be granted enough time to clear the tunnel and the prison grounds before Hobart would send through another group.

They intended to keep a running total of the escapees who went down the fireplace passageway so that their comrades could better conceal their absences the next morning. If fate favored the plotters and the tunnel's existence remained secret for another day, the same procedure might be followed on subsequent nights. However, no realistic system was adopted for regulating these later exoduses. Preoccupied with their own plans, the tunnelers gave little thought to what might happen after they and their companions left Libby.

* * *

At seven o'clock, the short winter day ended, and Richmond was wrapped in darkness, relieved only by the anemic light from a two-day-old moon. Conditions were nearly ideal for a breakout, and the tunnelers and their traveling companions were more than ready, though their nerves were frayed by adrenaline and anticipation.

They quietly assembled in the kitchen's shadows to receive Rose's final whispered instructions. Then, they opened up the fireplace and began dropping through the passage that Major Hamilton had begun more than seven weeks earlier.

One by one, the men whispered the password to Hobart, lowered themselves into the opening, and disappeared. While this tableau was being acted out, the second group of fifteen danced in the kitchen to a banjo reel to burn off their pent-up anxiety and draw attention away from the fireplace.

Rose went last, shaking hands with Hobart and quietly wishing him luck. Hobart closed the fireplace and marked the time. In one hour, he would reopen the entrance to admit the dancers, who were now "whooping and hollering, an all-male hoe-down," reported one observer.

* * *

On this night, the cellar's scuttling rats, awful stench, and oppressive darkness scarcely made an impression on the nervous men waiting at the tunnel entrance for Rose. The colonel lingered at the bottom of the fireplace chute until he was satisfied that Hobart had pulled up the ladder and replaced the bricks. Rose arranged his men by rank for the journey through the tunnel. Then he shook each man's hand, warmly thanked him, and bade him "Godspeed and farewell." As the highest-ranking officer, Colonel Rose entered the tunnel first, with Major Hamilton at his heels.

Some of Rose's fellow conspirators—the fanners, lookouts, and dirt removers—until this moment had never ventured far into the tunnel. They found it a tight fit. While the portal had been expanded to a roomy two and a half to three feet, the tunnel quickly narrowed to an average of 20 to 24 inches, and at one point to just 16 inches. After plunging at a shallow angle for a short distance, the bore leveled out and, more than 50 feet from the east cellar, it began angling steeply upward to its outlet several yards from the street. As the men wriggled forward on their bellies, "at some points there was considerable friction," one of them dryly noted.

The Confederates had boarded up the windows beside the kitchen fireplace to discourage communication between prisoners and guards, but Colonel Hobart was able to see Canal Street from the windows at the other end of the room. He was transfixed by the "living drama" that he witnessed. "Dancing in one part of the room, dark shadows disappearing in another part, and the same shadows appearing on the opposite walk."

Among the dancers were Captain Charles Earle and his traveling companion, Captain Charles Rowan. Earle had blundered into the kitchen one night while the diggers were descending through the fireplace and, on the spot, had become an honorary member of the escape cabal, although he was never assigned to a tunneling team. Equipped with a copied map of the peninsula and dressed in "traveling clothes," Earle and Rowan were waiting for Hobart to reopen the chute to the east cellar. Their haversack contained chunks of corn bread, scraps of meat, an empty bottle for water, and copies of the New Testament.

When an hour had passed, Hobart gave the signal and the second group of escapees stopped dancing and assembled at the fireplace. One by one, they vanished into the curving chute to the east cellar.

From the upstairs rooms, Hobart heard "a wild excitement and enthusiasm" as officers watching from the windows began to see their comrades walking down Canal Street. Hobart sealed the fireplace again to give the second flight sufficient time to leave the prison before others followed them through the tunnel.

Earle timed his transit from the fireplace to the tunnel exit: it took fifteen minutes. The trip through the tunnel itself lasted just two to three minutes, but Earle never forgot it. "No light and airy opening, but a narrow, dark, damp hole, just large enough for one to pull himself through." In the tunnel with him was a man who was "kicking and floundering against the walls," causing a deafening racket. The noise reminded Lieutenant Leander Williams of the 73rd Indiana of "wagons in a covered bridge."

*   *   *

Colonel Rose emerged from the tunnel and skirted the brick building where the Confederates had stored the five thousand undelivered prisoner boxes, tensely watching for Rebel looters. There were none. Reaching the arched wagon way, he quietly opened the gate and peered into Canal Street to determine the sentinel's location.

From the Lower Chickamauga Room, Rose had patiently watched the Canal Street guards for hours, and he knew exactly what to expect now. The archway lay a dozen feet east of the spot where the pacing sentry stopped at the end of his march, pivoted so that his back was to the gate, and began stepping off forty-five measured paces to the west. Rose believed that during this interval, a small group of men might slip through the archway and melt into the Canal Street shadows before the guard began marching eastward toward the gate again.

The sentinels on duty this night were not the usual ones that guarded Libby prison—two companies from the 18th Virginia Artillery that had fought at First Manassas. The regular guards were being given a rest, and soldiers from a brigade that was passing through Richmond were filling in for them. This substitution was a lucky break for the escapees.

Rose signaled that the way was clear, and stepped into the street. The others followed, two or three at a time, some of them having to wait until the guard again turned his back. The danger was exhilarating to men who, after being penned up in the foul warehouse rooms for months, were suddenly free to walk around outdoors. In this first group of escapees, Johnston, awed by his liberation from his private hell in the east cellar, observed, "We did not linger, and soon we were out of sight of the hated place."

A sentinel stationed across the street near the Kanawha Canal later reported having seen men walk out of the gateway. But he guessed that they were Confederate soldiers robbing the warehoused Yankee boxes and did nothing. Other Rebels, too, apparently reached the same erroneous conclusion; from his Libby room, Colonel Frederick Bartleson saw a guard stop and hesitate, but the guard issued no challenge as a group of escapees entered the street. Bartleson judged the guards' mistake to be "a righteous retribution for their mean pillaging of our boxes," and believed that "the hand of God was in it."

* * *

In the upstairs rooms, the whispers were rising to a crescendo as more captives, seemingly by magic, appeared on Canal Street and then vanished into the city. Lieutenant James Wells and other officers who knew about the escape plan got up from their sleeping places on the floor with their packed haversacks and overcoats, and headed for the kitchen. Rebuffed by Lieutenant Mead when he had tried to barter for a pair of Mead's boots, Wells now simply took them, and "like the Arab, folded my tent and silently stole away." On his way out of the room, Wells plucked a hat belonging to a West Virginia lieutenant and "hurriedly placed it where, in my own judgment at least, it would do the most good."

Men were pouring out of the upstairs rooms and down the stairway into the kitchen. Hobart struggled to control the crowd, which was nearly hysterical at the prospect of imminent liberty. He failed. Some of them opened the fireplace and began jumping down the chute "like sheep," one escapee wrote. Realizing that there was nothing more that he could do, Hobart let down the rope ladder and descended to the cellar with his companion, Colonel T. S. West—the last of the original thirty conspirators. The secret passage and the tunnel were now open to everyone in Libby Prison.

* * *

Amazingly, Lieutenant Frank Moran had slept through everything. A friend shook him awake around midnight to excitedly tell him about the mass escape. Joining the throng of murmuring men at the south windows, Moran watched one escapee after another appear briefly on the sidewalk across Canal Street from the wagon entryway before melting into the shadows. Although he was unprepared for a cross-country winter journey through enemy territory—he had no food, his shoes were dilapidated, his clothing was in tatters—Moran decided to escape. "I had not a crumb of food saved up . . . but as I was ill at the time, my appetite was feeble."

Lieutenant A. C. Roach, Colonel Streight's aide, found the spectacle to be "truly amusing and ludicrous," and was not tempted to join the exodus. Officers were "running hither and thither, begging, borrowing, stealing and buying a few crusts of corn bread, meat, or other edibles . . . anything to stay the cravings of nature for a day or two, or until they could find a friendly negro."

When Moran and his new traveling partner, Lieutenant Harry Wilcox, reached the kitchen, they beheld a surreal scene. As a sentinel outside stamped his feet against the cold a few yards from the boarded-up kitchen window, several hundred men silently struggled with one another in the murky room for choice spots in front of the fireplace, which was steadily swallowing men, one at a time. The suffocating pressure exerted by the desperate captives caused some of them to faint, while others "begged to be released." As Moran forced his way through the heaving mob toward the fireplace, someone on the edge of the crowd suddenly cried, "The guards, the guards!"

The prisoners swung around and raced for their sleeping quarters. To make the fireplace more accessible, they had moved aside the cook stoves and utensils, but now, in their hasty flight, they crashed into the pots, pans, skillets, and stovepipes that were strewn about the floor, "making such a tremendous noise that one might easily have mistaken it for the convulsion of an earthquake," wrote Captain David Caldwell. Unhinged by their fear of being caught, some men clambered over the heads and shoulders of their fellow officers, while others were swept off the unbanistered stairs onto the milling men below. Moran, one of the lightest men in the crowd, was knocked off his feet and briefly carried along by the stampeding men like a leaf, until he was dumped on the floor and trampled underfoot. Willard Glazier, too, was borne along in a wave of panicked captives that raced upstairs and through a doorway. It was, he wrote, "as though we had been shot from a cannon . . .

most of us not stopping until we struck the wall on the opposite side of the room."

Crouched among the stoves was Lieutenant Wells, in his stolen boots and hat. His partner had fled with the others, but Wells was waiting for the hubbub to subside. Hearing no approaching guards, he quietly slipped into the east cellar. An officer there led him to the tunnel, and he wriggled through it with his overcoat draped over his legs. When he emerged from the tunnel Wells breathed clean air for the first time in six months. "I felt the soft ground under my feet, and looked over and about me, as if to assure myself that it was not all a dream. I never felt a greater determination to accomplish a purpose in my life. . . . My senses were as alert and quick as those of a wild animal."

Bruised and disoriented, Moran struggled to his feet to find that he was alone in the kitchen. His companion, Lieutenant Wilcox, had disappeared. Moran cautiously approached the fireplace, his ears attuned to the tread of Rebel guards. Concluding that the alarm had been false, Moran removed his overcoat, dropped it through the fireplace portal, and then plunged feet first after it, landing in a heap in the cellar. His first sensation was of rats, "a troop of them," swarming over his hands and face. "Several times I put my hand on them, and once I flung one from my shoulder." Groping blindly in the cellar's darkness, his hand grasped a stave, with which he lashed out at the rodents. "I burned to be away," Moran wrote. But he had to find the tunnel first, and he did so only after a "long and horrible hunt" while stepping repeatedly on the scurrying rats, eliciting loud squeaks.

Moran knew that he had found the tunnel when his hands lighted upon a pair of heels that vanished at his touch. He pursued the feet through the opening. Emerging in the 30-degree Richmond night, he discovered that the man ahead of him was Lieutenant Charles Morgan of the 21st Wisconsin. Moran joined Morgan and his fellow Wisconsinite, Lieutenant William Watson, in their dash through the archway into Canal Street. They would stick together for the next several days.

* * *

Back in the sleeping rooms, the prisoners who had succumbed to the great panic tensely awaited the arrival of the Rebel guards. An hour passed, and there was no sign of the Confederates. The captives cautiously rose from their sleeping spots and returned to the kitchen, careful not to step on the scattered cookware. This time, there was no hysteria, no melée. The Yankee officers calmly queued up to take their turns dropping through the fireplace to the cellar. But no one kept a tally of the escapees so that their absence might be concealed the next morning.

* * *

In this large exodus of captives was Colonel Abel Streight, his generous epicure's figure surprisingly intact after nine months in Libby and three weeks in the dungeon. While many of his fellow officers had nearly starved, Streight had sustained his bulk through the bounty of his Northern friends, and the ministrations of the loyal subordinate officers who lived with him in "Streight's Room." Streight was Libby Prison's largest man at 6-foot-2 and 225 pounds, and even his aide, Lieutenant Roach, granted that his commander was "somewhat inclined to corpulency."

When it was Streight's turn to wriggle through the tunnel, he became stuck at the spot where it curved around a boulder and narrowed to 16 inches in diameter. For a few tense moments, the colonel could neither advance nor retreat, but after a struggle he was able to back out. In the cellar, Streight removed his overcoat, vest, and shirt and tied them to his legs before trying again. This time, he squeezed through, with his clothing trailing behind him.

* * *

As he waited in the cellar to begin his subterranean journey, Captain John Lewis was startled to see a Rebel sentry's legs framed in a barred window opening a few yards away. The guard was pacing his post on the prison's northern, Cary Street side. Lewis marveled at the Rebels' obliviousness to "what the Yankees were doing almost under his feet."

The breakout had been under way for several hours, scores of prisoners had entered Canal Street by twos and threes in plain sight of hundreds of their comrades, and had even been seen by sentries. Yet the Confederates were unaware of the historic escape that was occurring under their noses in their flagship war prison, less than a mile from the Confederate White House.

A city watchman extinguished the gas lamps on Canal Street at 1:30 AM on February 10, plunging the escapees' thoroughfare into complete darkness. In the Libby kitchen, in the east cellar, and at the wagon archway, activity quickened, for while the watchman's action better concealed their movements, it also heralded the coming of daytime, when it would be dangerous to be seen in Richmond. The escapees had to leave immediately if they were to reach the city's outskirts and slip through the Rebel defensive positions before the first light of day.

Around 3:30 AM, the chute through the kitchen fireplace was closed, rebricked, and smudged with soot, and the stoves and cookware were returned to their usual places. A preternatural quiet descended upon Libby Prison.

12

Flight

"I was overcome with terror so . . . that I quite lost my voice for some time. . . . How my heart ached all the while for their peril."

—Elizabeth Van Lew, meeting with escaped prisoners hiding in Richmond

"Our guards now think 'we are right smart.'"

—Union prisoner Robert Sneden, writing after the tunnel escape

A s THEY LEFT the gated archway, many of the escapees walked one block to Cary Street because they knew it would quickly take them out of east Richmond and into the countryside. The captives were shocked to find it brightly lit and thronged with soldiers and civilians. Cary Street was part of Richmond's Shockoe Bottom commercial district, and despite the hour and the cold, many shops were still open and busy. The crowds, noise, and lights were a rude assault on the officers' senses after their long months of isolation. Danger lay all around them.

It was providential that Libby Prison lay in east Richmond, because most of the fugitives were bound for General Benjamin Butler's forces near Williamsburg, 100 miles to the southeast. The escapees believed that if they negotiated Cary Street's gauntlet of bright lights and people, darkness would cloak them in the residential districts beyond.

Their first challenge was merging with the bustling foot traffic without drawing attention to themselves. Colonel Hobart pretended to be an elderly man with a consumptive cough, clinging to the arm of his companion, Colonel T. S. West. Lieutenant J. M. Wasson saluted groups of Rebel troops who were out on nighttime liberty until he was clear of the city. Captain W. S. B. Randall walked arm in arm with Lieutenant N. S. McKean, and they survived several close calls.

Captain I. N. Johnston and Lieutenant J. C. Fislar encountered two women who were conversing on the sidewalk outside an east Richmond home. The women turned and looked at them with frank curiosity. This was understandable, for Johnston was pale, gaunt, and filthy after spending nearly two weeks in "Rat Hell." The officers heard one of the women say that the pair looked like Yankees. "We did not stop to undeceive them," but walked on, Johnston reported. Guiding by the North Star, Johnston and Fislar slipped past the Rebel batteries and encampments that encircled the city. The frigid wind was an unexpected ally, stirring up piles of dead leaves that drowned out the sound of their footsteps.

After leaving behind the bright lights of the commercial district, Captain Charles Earle and his companions had the misfortune to fall in with a group of jocular Rebel soldiers. The escapees affected a casualness they certainly did not feel, "talking to ourselves on subjects similar to those we observed they were discussing." They were alarmed when the Rebels continued on the same eastward path as the escapees, evidently returning to a camp on the city's outskirts after an evening furlough. For nearly half an hour, Rebels and Yankees strolled together, until Earle and his comrades slowed to allow the Rebels behind them to pass, and then turned in another direction when they reached the Richmond & York River Railroad.

✳ ✳ ✳

As the eastern sky turned a pale, milky color, the fugitives sought hiding places to pass the daylight hours. Randall ruined his shoes fording the Chickahominy River, shredded his thin clothing crashing through the thick riverside brush, and finally bedded down in a thicket. Captain David Caldwell and Lieutenant W. A. Williams traveled three and a half miles before the approaching dawn drove them under a brush pile, where they spent the next twelve hours.

Not all of the escapees spent the night putting miles between themselves and Richmond. Colonel Abel Streight and three companions went into hiding in the city. Another band of escapees sought refuge in the home of Elizabeth Van Lew.

* * *

Van Lew had prepared a room for Libby's prisoners upon learning from Abby Green that an escape was planned, but she did not know when it would occur. The Confederate draft had recently snared her thirty-nine-year-old brother, John, who was as unwilling as his sister to serve the Rebel cause. And so John had deserted and, like the Libby refugees, he was dodging the Confederate authorities. The night of the Libby breakout, Elizabeth Van Lew put on a sunbonnet and plain clothing to mask her identity and went to the home where John was in hiding. She intended to help her brother slip out of Richmond the next day and flee to the North.

Early the next morning, Elizabeth Van Lew's driver brought disturbing news with the provisions for John's journey to the North: there had been a breakout from Libby Prison. Worse, Van Lew's servants, not knowing of the escape, had turned away strangers who had sought refuge at the mansion during the night, fearing that they were disguised Confederates who intended to entrap Van Lew. Van Lew's preparations for the escapees had been wasted, and they had to take their chances in the countryside. "We were greatly distressed," she wrote. "We were so ready for them, beds prepared for them."

Van Lew also realized that it was now too risky for John to leave Richmond, with soldiers stopping everyone who aroused suspicion. She had to act quickly. Putting on her best clothing and adopting her most persuasive manner, Van Lew paid a visit to General John Winder. Nearly two years earlier, she had persuaded Winder to permit her to visit the war prisons. Now she told Winder that her brother was seriously ill and unable to join his unit. Winder said that while he could not exempt her brother from the draft, he would change John's orders so that he could serve in a Richmond-based regiment.

* * *

LIBBY PRISON, FEBRUARY 10, 1864

After completing the morning head count, Erasmus Ross looked at his tally sheet and frowned. The Union officers watched with barely concealed glee; even the handful of captives who had slept through the escape now knew about it. Suspecting that the Yankees were playing games with him again, Ross sent a guard down to his office to retrieve the prisoner list, and began calling out each name. As the ranking officer, General Neal Dow's name was read first. He passed through the double file of guards beside the doorway and stood beside Ross. One by one, each captive's name was called, in order of descending rank. When all the names had been read, Ross's nerves began to tingle with an unpleasant premonition; the tally still did not jibe with the prison roster. He commenced a second roll call.

Four hours after the first head count, Ross completed the second roll call. General Dow, again standing next to Ross, was amused by the "expression of blank astonishment and despair" that stole over the clerk's face as he paged in mounting disbelief through the roster. A shocking number of the roughly twelve hundred men who were supposed to be present were in fact absent. "Why, it is a *hundred* and ten!" Ross exclaimed to himself, giving the impression that he had at first thought the count was off by ten prisoners. With that, Dow reported, Ross "turned deathly pale." A more careful enumeration revealed that there were 109 missing prisoners: twenty-one majors and colonels, thirty-five captains, and fifty-three lieutenants. Alarm spread through the Richmond war prison hierarchy.

* * *

The stunning news that 109 Union officers had escaped from Libby Prison spurred the Confederate command into furious activity. The prison security force was immediately doubled. Richard Turner galloped off on horseback with some Rebel guards to search for the fugitives. Confederate cavalry units began patrolling the roads north and east of Richmond, pickets were doubled on nearby roads and bridges, and all units operating in the area were ordered to keep an especially close watch on the 100 miles between Richmond and the Union lines at Williamsburg. Outside the military chain of command, fragmentary reports of the mass escape began reaching Richmond civilians. It would soon become a press sensation.

General John Winder ordered the arrest of the sentinels who had been on duty the previous night, suspecting their involvement in the escape. At the prison at Castle Thunder, the Rebel guards were searched for greenbacks and interrogated. To reinforce Winder's suspicions and to divert the Rebels from the tunnel, some captives pried the iron bars from an upstairs window and dangled a rope made of blanket strips to the ground.

Major Thomas Turner did not fall for the blanket ruse. Turner and his adjutant, Lieutenant John Latouche, who in civilian life had been a tailor and so had an eye for detail, conducted a minute examination of the prison grounds for more clues. The prisoners hadn't made their task easy: the fireplace had been sealed, the tunnel entrance had been stopped with a chunk of granite, and the exit was covered with a hogshead—the hope being that the "underground railroad," as some prisoners had nicknamed it, might be reused. "Probably one more night might have emptied the prison," wrote Colonel Streight's aide, Lieutenant A. C. Roach, who had stayed behind.

Under questioning, the Libby guards reported having seen men near the warehoused Yankee boxes, but told officials they had presumed that they were Rebel looters. Latouche returned to that area and carefully examined the ground near the tobacco shed. When he overturned the hogshead, he discovered the tunnel exit. Confederate soldiers converged on the site, and Latouche sent a servant boy down the hole. Minutes later, the boy's cries were heard in the east cellar. Colonel Rose's tunnel was no longer a secret.

General Winder ordered Richmond's church bells to sound a "general alarm." Guards were posted at both ends of the tunnel, and Rebels crowded into the cellar to see Rose's handiwork. Almost without exception, they expressed grudging respect for the Yankees' ingenuity. But it remained a mystery to the Confederates how the captives had gained access to the cellar.

As the day drew to a close, the guard was doubled at all of the Richmond prisons, and cavalry patrols roamed the streets. "Our guards now think 'we are right smart,'" Robert Sneden wrote with pride in his diary. Indeed, within days the tunnel would become known as the "Great Yankee Wonder."

✶ ✶ ✶

Most of Richmond found out about the escape from the newspapers. The headline in the February 11 *Daily Richmond Enquirer* shouted:

EXTRAORDINARY ESCAPE FROM THE LIBBY PRISON–
ONE HUNDRED OFFICERS HOMEWARD BOUND

On the same day, the *Daily Richmond Examiner* effused: "One of those *extraordinary* escapes of prisoners of war . . . occurred at the Libby Prison." The *Richmond Dispatch* described it as "the most important escape of Federal prisoners which has occurred during the war." The tunnel, reported the *Examiner*, "lay directly beneath the tread of three sentinels."

Both the *Examiner* and *Enquirer* speculated that the conspirators reached the cellar by boring through the floor of the prison hospital, after somehow breaking into the hospital from a second-floor room. The newspapers never explained how the prisoners might have done this without alerting the hospital's patients, orderlies, and doctors. The *Dispatch* simply reported that the captives had broken into the cellar "by some process" besides the sealed kitchen stairway.

The *Examiner* published a list of the "principal" escapees—the colonels and majors known to be missing. "Among them we regret to have to class the notorious Straight [*sic*]." The *Enquirer* named just one of the escapees, the "somewhat famous, or rather infamous, Col. A. D. Streight," and so did the *Dispatch*, reminding its readership that he was "a notorious character . . . charged with having raised a negro regiment." Although "the credit of the Confederacy" would suffer each time a captive reached the Union lines, citizens could be thankful that the "leak" was discovered before more prisoners could escape, wrote the *Examiner*. "The exodus would have been continued last night, and night after night, until there would have been no Yankees to guard."

* * *

While Richmond was reading about the escape, Rebel patrols were rounding up Yankee fugitives outside Richmond. Richard Turner's posse captured eight men on the roads. More were caught by Confederate pickets along the Chickahominy River. Among them was Lieutenant J. M. Wasson, who had jauntily saluted Confederate troops on his way out of Richmond. Wasson and his companions surrendered when five Rebel soldiers fired on them.

By nightfall of February 11, twenty-two men were in custody, including Lieutenant Frank Moran, who had impulsively set out without food or warm clothing. When Rebels caught up with Moran and lieutenants Charles Morgan and William Watson in a swamp north of Charlottesville, the Confederates pointed to smoke in the near distance and told the fugitives that it came from a Union outpost; they had almost made it. Placed in irons, they were locked in Libby's rat-infested dungeon. The ordeal had sapped Moran's vitality; ill when he escaped, he was now in far worse condition. Another escapee warned Richard Turner that Moran would probably die if kept in the cellar on a bread-and-water ration. Turner reportedly replied, "Well, damn him, let him die. That's what he's here for."

Lieutenant Melville Small of the 6th Maryland revealed the kitchen fireplace's secret to the Rebels after he was recaptured, clearing up the last major mystery about the escape. The Confederates now knew the essential details of the breakout, down to the false alarm in the kitchen and the order in which the first group had crawled through the tunnel.

The prison authorities and the Richmond newspapers got some of the details wrong, however. The Rebels, who had initially overstated the impact of Colonel Streight's raid, now identified him as the tunnel's mastermind. The *Enquirer* concluded that Captain Johnston was the principal digger on account of his unexplained absences, and reported that his tools were an old hinge and a sugar scoop stolen from the hospital. The *Enquirer* also reported that the tunnelers dumped the excavated dirt into a knapsack and dragged it from the tunnel for disposal.

The Southern newspaper reports about the escape and manhunt were slow to penetrate the North. The war had severed civilian telegraph service between Richmond and Washington. Nonmilitary correspondence from the South now reached Washington on foot and horseback, and by flag of truce boat. It was not until February 16, a week after the escape, that the *National Intelligencer* published its first article on the breakout, a short account reprinted from a Richmond newspaper.

* * *

The first escapees entered the Union lines near Williamsburg on February 14, after traveling 100 miles in five days. Cold, hungry, and wet, the two fugitives had been pursued by patrols and dogs during their flight through fields, woods, and swamps. They were the vanguard of twenty-six prisoners who arrived in Williamsburg that day.

Their arrival was a clarion call to action. "My cavalry are in motion, scouring the Peninsula to cover the escape of the rest," General

Isaac Wistar reported to General Benjamin Butler's chief of staff, Colonel J. W. Shaffer. "Several colonels, among them Colonel Streight, are on the road, but the path is hard." Wistar sent fresh cavalrymen up the peninsula each day to search for fugitives, alternately dispatching the 1st New York Rifles and the 11th Pennsylvania Cavalry. They patrolled with tall guidons to heighten their visibility. A Union gunboat cruised the James and Chickahominy rivers, its crew scanning the riverbanks for gaunt, ragged men. On February 15, nine more escapees reached safety.

Wistar believed it was imperative that the escapees quickly cross the Chickahominy River to have a realistic chance of success. Even as he made strenuous efforts to find the fugitives, he privately did not think many of them would reach the Union lines, writing, "If one-fourth the escaped prisoners get in it will surprise me."

* * *

It was an immutable fact that anyone traveling from Richmond to Williamsburg had to somewhere cross the Chickahominy River. Flowing out of the Virginia Piedmont, the Chickahominy passes north of Richmond, and turns southeastward to flow down the peninsula, before veering south and emptying into the southeast-flowing James River outside Williamsburg. The rivers enclose a boot-shaped area between Richmond and their confluence, with the Chickahominy representing the boot's uppers, and the James its sole. In peacetime, one simply rode down the Charles City Road toward Williamsburg and crossed the Chickahominy on a ferry at the boot's toe. But in February 1864, the Confederate Army controlled the area confined by the two rivers. Butler's forces only occasionally contested Rebel control of this region, which was alive with mounted enemy patrols.

But north and east of the Chickahominy, Wistar's cavalry roved the countryside, facing token opposition from a small Rebel scout unit. The captives who had studied maps of the area, or who had fought on the peninsula in 1862, knew that they should cross the Chickahominy immediately northeast of Richmond, before it became swift and deep. Two men who escaped through the Libby tunnel drowned while trying to cross the river 15 miles east of Richmond, near Grapevine Bridge.

✳ ✳ ✳

Dogs were as much a threat to the fugitives as were the Confederate patrols. On his first day of freedom, Captain David Caldwell, the temporary chaplain, was hiding in a brush pile with Lieutenant W. A. Williams and watching a man collect firewood, when the man's dog picked up their scent and began following it, straight to where they lay. Caldwell and Williams offered up a silent prayer for a reprieve and, to their relief, the dog suddenly ran off in another direction, distracted by something more interesting. Lieutenant Morton Tower of the 13th Massachusetts and Captain George Davis of the 4th Maine killed a baying hound with a cudgel and a pocketknife. Colonels Hobart and West walked backward across roads and jumped over brooks to confound their canine pursuers. Lieutenant Albert Wallber of the 26th Wisconsin and his companions had followed the Richmond & York River Railroad tracks out of Richmond to a hiding place in the woods when they heard a train horn being blown repeatedly. As they watched from a distance, a farmer appeared with two bloodhounds, which immediately got onto the fugitives' trail. The prisoners ran, crisscrossing their paths to confuse the tracking dogs, and were able to leave behind the hounds and their handlers.

Caldwell's luck continued to hold when, just as he and Williams were setting out from their hiding spot, they spotted a Rebel squad that was forming a picket line. As the fugitives began running away, a whirlwind sprang up between them and the Confederates, screening their getaway. After nightfall, the fugitives stripped off their clothing and, trembling violently in the wintry air, waded naked across the freezing Chickahominy River, with the water reaching their armpits.

Snow began falling the next morning, and it continued all day. Caldwell and Williams elected to continue traveling, reasoning that the snowfall concealed their movements, and the wet ground made sleep impossible anyway. The storm persisted into the night, and they slogged on, trying to ignore their fatigue, hunger, and extreme discomfort. Because the low clouds obliterated their view of their lodestar, the North Star, they grew increasingly uneasy, wondering if instead of heading toward Williamsburg they had gotten turned around and were walking back to Richmond. But they could not stop and risk freezing to death. "We must travel, even if we go back to Richmond, or else we shall surely perish," Caldwell told Williams.

Just when it seemed that they would succumb to the cold, they stumbled upon an abandoned slave cabin. They started a fire and were beginning to warm up when flames suddenly blossomed from the chimney and threatened to engulf the cabin. After a furious effort, they extinguished it with snow, and resumed their frigid trek. But then they met an elderly black man who led them past Rebel pickets to the cabin of a black woman, who prepared them a meal of ham, eggs, "hoecake," coffee, and buttermilk. She cleaned their muddy trousers and boots, and later guided them to a lightly traveled road leading to Williamsburg. After eight days and nine nights on the run, Caldwell and Williams finally met Union pickets.

* * *

The peninsular slaves had begun watching for the Yankees on the roads and in the fields and woods since learning about the escape from other slaves and from their masters' conversations. The escapees knew that they could count on the rural black population to give them a hot meal and a warm bed, or to lead them to the next safe haven. Knowing as they did that the war was being fought over the "peculiar institution" that kept them in bondage, and that a Northern triumph meant their emancipation, many slaves and free blacks aided the Libby fugitives, sometimes at great peril.

During their fourth night at large, Captain Johnston and Lieutenant Fislar encountered a black man who, after they had satisfied themselves as to his trustworthiness, took them to a house where other blacks shared their meal with the fugitives. When they departed, their guide escorted them for another four miles, leaving them with directions to Williamsburg. They encountered Union cavalry the next day.

A slave brought bacon and warm corn bread to the hiding place of Lieutenant Wallber and his comrades, and then handed them on to a free black named George Washington. Washington proved to be as brave and reliable as his name promised, rowing the prisoners eight miles down the Chickahominy and landing them on a riverbank beyond the Rebel pickets.

Colonel W. P. Kendrick of the 3rd West Tennessee Cavalry met a slave woman in a field who warned him to avoid her Confederate mistress. Weakened by hunger, Kendrick chose to ignore the warning, in the hope of obtaining food. The story that he and his companions concocted—that they were Confederates who had escaped *Yankee* custody near Norfolk—was so convincing that the Rebel matron, after feeding them, told them where the nearby Yankee units were located so that they could avoid them. Kendrick and his comrades followed the woman's directions straight to the Union forces.

Slaves and free blacks risked instant punishment, even death, whenever they aided the Yankee fugitives, and yet they lent assistance time and again. General Benjamin Butler later sent a $50 reward to a free black named John Caffey for having guided fugitive officers. The payment was the equivalent of a civilian medal for valorous service. "We hope [his services] will be extended in the future to any of our men whom he may meet with who need aid," Butler wrote.

Harper's Weekly praised black Americans everywhere for "helping our soldiers that escape from rebel prisons, and going from our midst to help them fight our battles for us. Hunted to death by the mobs in our cities, they retaliate, by joining our armies, and they do their duty on the battle-field."

* * *

Lieutenant James Wells, in his stolen boots and hat, had left Richmond in a hurry and waded across the Chickahominy River that night. The next day he hid in a thicket as a Rebel cavalry unit clattered by on a nearby road, looking for escaped prisoners. Wells followed the roads all that night. At one point, unsure exactly where he was, he climbed a signpost and, while holding onto it with one hand, lit a match. He was 12 miles from Richmond.

When he came unexpectedly upon an encampment of Confederate teamsters, Wells pretended to be one of them, "bursting out in the vernacular common to the class" and kicking a mule. The act evidently was convincing. Wells left the camp unmolested, reaching Williamsburg after a long hike through freezing rain and sleet. Along the way, he met Lieutenant Nineoch McKean of the 21st Illinois and Captain W. S. B. Randall, the tunnel digger.

Randall became separated from Wells and McKean while they were running from Rebel cavalrymen, and Wells assumed that Randall had been captured or killed. In fact, he was alive, though he suffered greatly from the cold, lack of food, and the chronic semistarvation that had reduced him to little more than a skeleton. Moreover, while eluding the Confederate cavalry, Randall cut his feet, they got wet and, before long, he had no sensation in them. Then bloodhounds got on his trail and began to close in. But unlike the other fugitives, whose only defense against dogs was prayer or violence, Randall had prepared for such an emergency; he had brought cayenne pepper. It had arrived in a box from home, hidden in a roll of butter. He sprinkled it liberally into his tracks, raced ahead, and found a hiding place where he could observe whether his handiwork was successful. Indeed, when the bloodhounds inhaled the pepper, they were convulsed with sneezes and gave up the pursuit.

Later, from beneath a tangle of vegetation near the Williamsburg Road, Randall watched a cavalry patrol ride up and down the road, searching for fugitives. The men wore blue, but Randall had not come this far to be recaptured because of carelessness. They might be Rebels wearing Yankee uniforms, a not uncommon sight in Richmond. So he spent the day patiently spying on the horsemen. At last, the men rode so close to Randall's hiding spot that he could see the initials "U.S." on their belts. He emerged from his blind into the open, "greatly to their surprise and my joy. . . . This certainly seemed to be the crowning event of my life."

And it literally saved his life: Randall had entered Libby Prison at 160 pounds, but he now weighed no more than 90 pounds. Randall and the other newly liberated men were feted at a procession of banquets, and smothered with kindnesses large and small, as compensation for the hardships that they had endured as captives and fugitives.

* * *

Of all the officers who escaped through the Libby Prison tunnel, none was as avidly sought as Colonel Abel Streight. A Unionist who visited Libby Prison on the pretense of obtaining more information about the breakout reported that the Confederates were "very lugubrious" over the fact that Streight remained at large, and were obsessed with catching him. "They would be content for all the other prisoners to get through safely, if they only could get the Colonel again in their clutches, either alive or dead," the Unionist wrote. But preferably dead: orders had been issued to kill Streight on sight.

Wild rumors circulated about Streight's purported whereabouts, and tips sometimes stampeded the Rebels into acting rashly. One day, government detectives raided a building on Richmond's Main Street. "I want to see Streight!" the *Daily Richmond Examiner* quoted a detective as shouting when his squad invaded a dentist's office and living quarters. Outside, the street filled with hundreds of people hoping to witness the capture of the raider colonel, just as the detectives flushed several men from a card game in another part of the building. One of them took to the rooftops with a detective in pursuit. The detective opened fire and the bullet struck a chimney. Some of the people in the street, mistaking another fleeing cardplayer for Streight, blazed away at him with pistols without hitting him. When the dentist poked his head from a window to see what the commotion was about, one of the on-lookers took a potshot at him, too.

∗ ∗ ∗

Colonel Streight, in fact, was still in Richmond. While their comrades dodged Rebel patrols and bloodhounds up and down the frost-rimed peninsula, Streight and his three companions were snug and well-fed in a home on the outskirts of the capital.

Days before the tunnel's completion, Abby Green had sent Streight instructions for contacting her if he escaped. Green's servant and Robert Ford, Richard Turner's black hostler, were the likely interme-diaries; they had shuttled messages between Green and Streight in the days preceding the failed Great Escape Plot.

After wriggling through the tunnel, Streight, Major B. B. McDon-ald, Captain William Scearce, and Lieutenant John Sterling followed Green's directions to a house near Libby Prison. There, a black woman led them to Green, who guided them to a working-class home belong-ing to Unionist John Quarles on Richmond's outskirts. A woman named Lucy Rice cooked and cared for the fugitives.

∗ ∗ ∗

Six days after the breakout, Rice arrived in her carriage at the Van Lew mansion. The escapees, she explained, wished to meet the leg-endary spy who had risked so much to aid the Union captives. Eliza-beth Van Lew was just as eager to meet them. She rode with Rice and her driver to the hideout, where the fugitives were waiting for her in the parlor.

The sight of Colonel Streight and his companions at first paralyzed Van Lew. "I was overcome with terror so . . . that I quite lost my voice for some time." Two of the men were clearly ill, but Streight appeared to be in good health. "How my heart ached all the while for their peril." When Van Lew was able to speak, she warned Streight that the Rebels harbored "particular enmity" toward him because they believed he had commanded a black regiment. He replied, "I did not, but would have had no objection." When Streight asked Van Lew to state her opinion on the war's cause, she was tongue-tied. "I tried to say, Democracy, though in my head I thought it was slavery."

She found McDonald to be particularly charming: "Broad-shouldered and kind-hearted, an honest, true, genial-looking man." McDonald placed in Van Lew's hand the chisel that had made the tunnel. When she suggested that he leave it with her, McDonald good-naturedly replied that he preferred to keep it. "We had a little laughing and talking and then I said good bye, with the most fervent, 'God bless you,' in my heart towards all of them."

Within hours, Confederate detectives descended on the Quarles home, possibly tipped off by the carriage driver. But Rice had anticipated trouble: when the policemen arrived, the fugitives were nowhere to be found. The detectives left empty-handed. They began watching Van Lew's home, but they did not raid it. Her social standing continued to shield her from interrogation and arrest.

Troubled by the inadequacy of her reply to Streight's unexpected question about the war's cause, Elizabeth Van Lew wrote a note to Streight. In it, she explained that she had meant to reply, the "slave power." The South had gone to war to safeguard the plantation owners' slave holdings, she believed, and not to preserve democracy. But the messenger to whom she had entrusted the note returned it to her. The men were gone, he said.

* * *

On the day that Van Lew visited Streight and his companions, General Isaac Wistar was acting on a request by Streight's friends. "Colonel Streight is concealed in Richmond, but at large," Wistar wrote to Colonel J. W. Shaffer, General Butler's chief of staff. "His friends desire the papers to state his successful arrival here, for obvious reasons. Please arrange it immediately with the Associated Press agent."

Streight's friends believed that a fictitious report of the colonel's safe arrival would throw the Rebels off his trail, creating an opportunity for him to at last slip out of Richmond. While the Confederacy was loosening its dragnet for the other escapees, Streight was still being actively sought in Richmond and in the countryside, although even the press was beginning to concede that he had probably gotten away. "As for Straight [*sic*]," wrote the *Daily Richmond Examiner*, "the Confederacy got more than ten times his value when it received [General John] Morgan back, and can afford to let him run. *Bon voyage*, whiskey *Straight* [*sic*]!"

The false story that Colonel Shaffer planted with the AP appeared in the *National Intelligencer* and other Northern newspapers. It was quickly followed by a *Richmond Sentinel* report that Streight and his companions had reached Fortress Monroe. The Confederates suspended their search for him.

* * *

Eight days after he wriggled half-naked through the Libby tunnel, the moment had finally arrived for Streight and his companions to leave Richmond. It was the coldest day of an unusually cold Virginia winter—7 degrees at daybreak and only 12 degrees at 2 PM. "Intensely cold," Van Lew wrote in her journal, after learning that the four men had left Lucy Rice's care that day, "so cold that some fellows in the Confederate service froze to death."

Carrying weapons and several days' food, Streight, McDonald, Scearce, Sterling, and a Rebel deserter who had agreed to guide them began walking north on the night of February 17. At the Chickahominy River, the deserter lost his nerve and turned back. The others pressed on, avoiding the roads and traveling at night, when darkness cloaked their movements—and when the danger of freezing to death was the greatest, for they wore only light clothing and had no blankets.

The fugitives elected to continue traveling north toward Washington, rather than turning southeast to Williamsburg, although the northern route was laced with rivers that would have to be crossed. At the Pamunkey, they built a raft and forced their way through a jam of ice floes. When they landed on the other bank, they attracted the attention of a Rebel guard, but they eluded him by hiding in a thicket. At the Mattaponi, they found a boat to carry them over and managed to build a fire on the opposite bank and bury their feet in the warm sand.

Miles from the mighty Rappahannock, they began to hear the rumble and crash of massive ice floats. The ominous sound distracted them from their fatigue and the numbness in their hands and feet. But before they could reach the south bank and begin puzzling out a way to reach the north bank, they were spotted. Rebel troops, citizens, and hounds set after them. The dogs, racing ahead of the others, overtook the fugitives, but they plied the canines with food scraps and urged them on, pretending that the objects of the manhunt lay farther ahead.

The ruse did not throw off the escapees' human pursuers. Two fugitives who were nearing collapse had to be supported by the others, and the Rebels closed in. Three times they were surrounded, but the officers evaded capture by hiding in thickets until nightfall. Darkness, however, brought no relief. Instead, hundreds of Confederate troops converged on the area to beat the bushes for their Yankee "game." Forming lines, they sealed off a neck of land hemmed by the Rappahannock on the north, east, and west, trapping Streight and his companions. Escape appeared impossible.

As the fugitives desperately sought a way out of the trap—their nerves near breaking, their limbs trembling with fatigue, exhaling the frosty air in puffy white clouds—Virginia's slaves came to their rescue. They hid the Yankees in a cabin on the plantation where they worked, and then made a show of joining the hunt for the men they had just concealed. The perplexed Rebels lost the fugitives' trail and broke off the search. When the troops left the area, the slaves rowed the men across the Rappahannock.

The officers resumed their flight to the North, hiking fifty miles to the south bank of the Potomac. There a black man led them to the home of a Union sympathizer, who lent them his boat. But when they tried to cross the river, Rebel pickets opened fire on them from shore, wounding one of the fugitives. The officers returned to Virginia.

Two days later, on February 28, they again set out in the Unionist's boat, and this time succeeded in reaching Blackstone Island. The steamer USS *Ella* of the Potomac Flotilla picked them up the next day and took them to Washington.

On March 1, the *National Intelligencer* reported that Streight, McDonald, Scearce, and Sterling had reached the capital. Washington would soon hear the story that Streight had to tell.

* * *

Of the 109 officers who escaped through the tunnel, fifty-nine reached safety, two drowned, and forty-eight were recaptured. Captain Junius Gates of the 33rd Ohio was the only one apprehended in Richmond. Amazingly, the Rebels killed none of the fugitives, and wounded just one of them.

Two officers suffered the embarrassment of being captured by a pair of boys on horseback. The teenagers ordered the prisoners to march ahead of them to one boy's home, where they were turned over to the militia. The fugitives were "very much chagrined" when they learned that their captors were neither armed, nor cavalrymen, nor even in the Rebel army. In its gleeful recounting of the episode, the *Richmond Dispatch* identified the intrepid boys as a drugstore employee and a farmer's son.

After crossing the Chickahominy with four companions, Captain John Lewis became ill and was left behind. Richard Turner and his posse found him and took him into custody. Evidently because Lewis had well-connected Confederate relatives, Libby's despised warden displayed surprising magnanimity. Turner lent Lewis a horse to ride to Richmond and bought him a drink before taking him to the Libby hospital.

Still, the recaptured escapees experienced the full measure of dejection and despair when they were promptly shut away in Rat Hell. Some were chained and some were not, but they all wore the same stained, ragged clothing that they had on when they wriggled through the tunnel. Their comrades upstairs tried to cheer them up by slipping them food through the kitchen floorboards and offering them encouraging words and mordant humor. "There's no use in praying down there, boys," a prisoner jibed as he removed a floorboard and began handing down food. The other captives were living on meager rations themselves, but they shared the little that they received.

✳   ✳   ✳

Lieutenant Latouche's discovery of the tunnel vindicated the Libby guards; they were freed and returned to duty. As the manhunt wound down, and General John Winder and Major Thomas Turner turned their attention to other matters, the Libby staff breathed easier. Moreover, many of the Richmond prisoners would soon be transferred to war prisons in the Deep South.

There was a last matter to settle before General Winder closed the book on the tunnel escape: the issue of Richard Turner's hostler, Robert Ford. Winder's plug-uglies had learned about Ford's role as an intermediary between the prisoners and the Richmond Unionists. The prison commander chose to make an example of Ford, so that other trusties would not be tempted to aid the enemy. Winder sentenced Ford to five hundred lashes—an extreme punishment. Before the U.S. Navy abolished floggings in 1850, a seaman rarely received even one hundred lashes.

The sentence was carried out in Libby's cellar. Four men held Ford over a barrel, and Richard Turner administered the punishment—every single lash of it, as Ford's cries echoed through the prison. Mercifully, Ford lapsed into unconsciousness before Turner finished. The flogging would have killed many men, but Ford survived, although he was incapacitated for six weeks. Before he was fully recovered, Ford managed to escape from Libby and disappear into Virginia's black community. Two months later, he crossed the Potomac River and returned home.

* * *

As they awaited the denouement of the Confederate manhunt for their comrades, the captive Union officers endlessly parsed and debated the tunnelers' amazing feat. It was now clear that Colonel Thomas Rose was the principal figure behind the tunnel's success. As Colonel Frederick Bartleson observed in his diary, it was Rose who walked the floor, hour after hour, pondering escape strategies while his comrades read or slept, and who stuck to the plan when others wanted to give up. The prisoners agreed that Rose, more than any other escapee, deserved to reach the sanctuary of the Union lines. They waited for the news of his safe arrival.

Rose was traveling alone through the woods and fields southeast of Richmond. He and Major Hamilton had left Libby Prison together, but two blocks away a Rebel hospital guard had stopped Rose and led him away for questioning. Believing that Rose had been recaptured, Hamilton proceeded solitarily to the countryside, and entered the swamps along the Chickahominy River, reasoning that Confederate patrols would be reluctant to venture into the wet tangle of trees, thickets, and fallen logs. Hamilton never left the marshes, traveling for seven nights in ice and water that sometimes reached his waist until safely arriving in Williamsburg.

The Confederate guard released Rose after a half hour of inventive explanations. Rose followed the Richmond & York River Railroad to the Chickahominy River, but he could not cross there. Confederates guarded the railroad bridge, and enemy cavalrymen were camped nearby. With the approach of dawn, Rose, feeling conspicuous in his blue uniform, crawled into a hollow sycamore log.

The next night, while wading across the Chickahominy he stepped into a hole in the river bottom and was drenched from head to toe. His wet clothing froze into a stiff tapestry. Although he forced himself to keep moving in an attempt to get warm, it did little good. With daylight, he curled up in a hiding place, his teeth chattering so much that he was unable to sleep. And when he had partly dried out, he got another soaking in a swamp while evading a Rebel patrol.

The temperature did not climb above single digits in the crystalline early mornings, and icy pains shot through the foot that Rose had once broken. Each step that he took was excruciating. Exhausted, hypothermic, and in real danger of freezing to death, Rose risked building a fire in a hollow in the middle of a cedar grove. The fire's delicious warmth lulled him to sleep. He awakened in pain of a different sort: the fire had burned his boot legs and charred part of his uniform; only the ice that coated the rest of his clothing saved it from the same fate.

On the third night, a Rebel cavalryman spotted him as he crossed an open area, minutes after Rose had slipped past a Confederate picket. The horseman galloped up to Rose and, seeing his gray cap, saluted him. Was he with the New Kent Cavalry? the man inquired. Yes, Rose replied without hesitation, and the cavalryman turned back. But his report evidently did not satisfy a superior officer, because a squad of Confederates came looking for Rose. He ran into a densely wooded area, where he dodged another Confederate picket.

The woods were obviously unsafe, but the only way out of them was to cross a half-mile open field. His situation seemed hopeless until Rose noticed that a ditch bisected the open area; there was a small chance that he might shake his pursuers if he could use it for concealment. Rose scrambled into the ditch and began crawling on his hands and knees. Briars and stones cut his hands, and with Confederates all around, the journey seemed to take forever. But he gained the other side of the field without being seen. Sprinting across a road, he plunged into another wooded area, and from a hiding place there he watched the enemy squad milling in apparent confusion before abandoning its search.

On the fifth day, Rose was just a few miles from Williamsburg when he saw Union troops approaching from the east. Believing that his ordeal was at an end, he sat down to wait for them, feeling what must have been enormous relief.

He was startled by a noise behind him. Turning around, Rose saw three blue-clad soldiers walking toward him from the opposite direction of the Yankees on the road. Presuming that they were the approaching soldiers' vanguard, he rose to meet them. When he was 100 yards away from the three, one of them shouted a challenge—and then the soldier, looking past Rose, stopped in his tracks and stared in obvious alarm at the Union troops on the road.

Too late, Rose realized that the three were Confederates in Yankee uniforms. He was caught.

A Rebel officer assigned one of his men to march the new prisoner to Richmond. However, Rose was not ready to give up. As soon as he and his escort were out of the sight of the other Confederates, Rose pounced on the guard, wrenched the musket from his hands, fired the round in the breech into the air, and flung it to the ground. He sprinted toward the Union troops, hoping that the gunshot had attracted their attention.

But before Rose had traveled more than a few yards, a dozen Rebel soldiers suddenly loomed before him, blocking his way. They clubbed him to the ground with the butts of their muskets.

This time placed under a strong guard, the master tunneler was marched to Richmond and cast into Libby Prison's dungeon with the other unlucky ones. Throughout Rose's thirty-eight-day confinement, Libby officials, deeply embarrassed by the tunnel escape, grudgingly doled out to him a bare daily subsistence of corn bread and water.

13

Fallout

"We will cross the James River into Richmond, destroying the bridges after us and exhorting the released prisoners to destroy and burn the hateful city; and do not allow the rebel leader Davis and his traitorous crew to escape."

—Colonel Ulric Dahlgren's instructions to his men

MAJOR THOMAS TURNER shook up his security detail at Libby Prison. Before the escape, the guards had rarely entered the captives' rooms after lights out, but now they did so at two-hour intervals, checking every fireplace and window, and every corner in every room. It greatly annoyed the prisoners, who had trouble enough sleeping on the cold, hard floors.

At any hour, the guards might also order the captives into the kitchen for a roll call. "We are counted and recounted, from morning until night," grumbled Colonel Federico Cavada. Pity the late-arriving prisoner; his punishment was to stand at attention under guard all night. The groggy men sometimes were required to remain in ranks for hours while the Rebels ransacked the upstairs rooms, ostensibly searching for knives, pistols, files, and excavation tools, but in reality "stealing whatever articles they fancied," complained Captain Albert Heffley. "They fear another underground railroad. . . . Nearly every day they add some new insults and indignities upon us."

Workmen installed a retractable stairway between the upper rooms and the kitchen, and at night raised it with a pulley to prevent the prisoners from entering the kitchen. The east cellar walls were sealed with concrete to frustrate any attempt to duplicate Colonel Rose's feat, and to collaterally diminish the rat swarms. Rebel sentries now visited the premises frequently. Throughout the prison, the barred windows were reinforced, the prisoners were warned anew not to lean on the iron bars or sashes—they would be shot if they did—and they were forbidden to throw bread to the beggar children who appeared below the prison windows each day.

The guards were quicker to shoot than ever. An Ohio adjutant was shot in the chest and killed as he was reading a book 10 feet from a window. No prisoners mourned when a Rebel guard on duty inside Libby carelessly poked his head out of a window to speak to another sentry and was instantly shot dead. Observed one prisoner with satisfaction: "So their hellishness begins on their own head. . . ."

The Libby Prison guard detail acquired Nero, a huge, ferocious-looking dog belonging to Castle Thunder's commandant, Captain George Alexander. Nero's handlers took him around the building foundations to sniff for secret earthworks. The prisoners were impressed by the dog's trained nose, conceding, "There would have been no chance of an undiscovered tunnel while that dog was within reach." But after further observation, they also concluded that Nero was not nearly as fierce as he looked. (Nero was later captured and ignobly sold at an auction on the steps of the Astor House in New York City.)

The strict security measures, together with Major Turner's earlier order restricting the prisoners' letters to six lines, had the effect of making them feel even more isolated and abandoned. And then, ten days after the tunnel escape, Confederate officials staged a public execution in the yard at Castle Thunder.

Spencer Deaton of the 10th East Tennessee—a Union Army unit from secessionist Tennessee's eastern mountains—was arrested in August 1863 for recruiting his neighbors. The Rebels denounced Deaton as a traitor and sentenced him to die. While awaiting execution in Castle Thunder, Deaton stopped eating, in the hope that he would die before he was hanged. The Confederates, however, labored to keep him alive, mainly by plying him with laudanum, and so Deaton survived to see the day of his execution, although guards had to carry him to the gallows and hold him up while the hangman did his work. Even Richmond's newspapers took no pleasure in the dismal proceeding.

* * *

In Washington, more than a dozen of the Libby Prison escapees, clad in new clothing and beginning to regain some of the weight they had lost in captivity, were busy describing to anyone who would listen how the Rebels mistreated their war prisoners.

In a letter to President Lincoln, Colonel W. P. Kendrick faithfully enumerated the abuses at Libby Prison, adding that Union enlisted prisoners on Belle Isle were also treated "brutally, cruelly. Many have frozen this winter; many more have died from actual starvation." Kendrick urged Lincoln to act to alleviate the captives' suffering.

After listening to Colonel William McCreery's account, General Benjamin Butler arranged for McCreery to meet with War Secretary Edwin Stanton in Washington. McCreery will "give you such information about our prisoners' fare and treatment as will demonstrate the necessity of retaliation if I do not succeed in starting the exchange, which I hope to do," Butler told Stanton.

Colonel Harrison Hobart, the fireplace gatekeeper, proposed to General Ethan Allen Hitchcock a limited exchange of one hundred to three hundred Union officers at City Point. Convinced by Hobart that the Confederacy would consent to a one-time prisoner swap—and, if it did not, that its refusal would greatly damage its image—Hitchcock decided to proceed.

Lieutenant Morton Tower believed that he had left his troubles in Richmond, until he encountered the immovable Washington bureaucracy, which was indifferent to his need for back pay so that he could begin his thirty-day furlough. On the third consecutive morning that he presented himself at the Treasury Department and was once again rebuffed, Tower remarked aloud, "It was might [*sic*] hard for a man who had just spent eight months in Libby Prison and with a 30-day leave of absence in his pocket, could not get the where-with-all to get home."

As Tower said this, a kind-looking man standing beside him turned to Tower and asked him for his name, rank, and regiment. "Wait a minute," the man said and left. When he returned, the man handed Tower a check containing his back pay, and his card. "On it, I found inscribed the name of Walt Whitman, known as the poet and soldier's friend." Whitman, already celebrated for *Leaves of Grass*, was a clerk at the Treasury Department. He was also a volunteer at a military hospital where, he wrote in "The Wound-Dresser,"

Bearing the bandages, water and sponge,
Straight and swift to my wounded I go.

* * *

No escapee made a greater impression in Washington than Colonel Abel Streight. Many of the others had already written their letters to President Lincoln, lobbied Congress to take action, and left on their furloughs when Streight and his three companions arrived in the U.S. capital on February 29. It didn't matter that he arrived late, because Streight was a celebrity.

To shield his Richmond benefactors from retribution, Streight fabricated an escape account for publication that omitted the Unionists' role and, in fact, everything that had actually happened to him. The *New York Herald* reported that Streight hid in the woods during a twelve-day odyssey to Williamsburg—not Washington—and described his party's hardships and brushes with the enemy in the eastern Virginia countryside. None of it was true.

Streight sat down at a desk in Willard's Hotel and composed a report on the Richmond prisons for Michigan congressman F. W. Kellogg of the House Committee on Military Affairs. "It is impossible for me to give you an account of all the acts of barbarity, inhumanity, and bad faith I have witnessed during my captivity," he began, and then gave vivid examples of each, beginning with his capture in Georgia and ending with his escape.

But bad as things were in Libby Prison, conditions were worse on Belle Isle, where Streight had seen hundreds of ragged, dying men being carried to the island hospital, and then left outdoors for days. "I heard one of the rebel surgeons in charge say that there were over twenty of our men who would have to suffer amputation from the effects of the frost." Streight bitterly concluded that the Confederates "seem lost to every principle of humanity, and it is my candid opinion that their brutality to our prisoners is only measured by their fears."

In response to Streight's report, the *New York Times* wrote, "Nothing in all the records of modern prisons is so brutally and savagely cruel, as the treatment of our officers and men in the Libby Prison by these Richmond authorities." It was as though the Confederate government, "fearing the opinion of the world, and not venturing on a cold-blooded massacre," had instead adopted "the secret and more sure method of unhealthy quarters and bad food—of typhus and starvation." While retaliation was "a terrible thing," the *Times* in effect advocated it, observing that the "slow wasting life of our brethren and friends . . . is a worse thing."

Streight's report and the newspaper accounts of the Libby escapees contributed to a rising groundswell of outrage over the Rebels' treatment of war prisoners. Feeling the public pressure, Congress appointed a Joint Select Committee on the Conduct of the War to investigate conditions in the prisons. At the same time, the United States Sanitary Commission began its own inquiry.

While he had the attention of the Union command, Streight did not pass up the opportunity to exact personal vengeance. Colonel James Sanderson, Libby's former food czar, had collaborated with the enemy, Streight charged, and had leaked details of the Great Escape Plot to the Rebels. Streight had not forgotten how the Rebels, to punish him for being a leader of the intrigue, had lured him into a sham escape attempt with the intention of killing him.

And so, when Sanderson was exchanged on March 7 and reached Washington, he found himself under house arrest, and then summarily dismissed from the army. His appeals for a hearing were ignored.

Strenuous pleading eventually won him the right to read the charges preferred against him by Streight, General Neal Dow, and others: abusing Belle Isle inmates while distributing provisions there; betraying the Great Escape Plot; speaking against the Union; and, while serving as Libby's culinary director, stealing a case of eggnog.

Sanderson determinedly lobbied the Union command for a hearing and, when he was at last granted one, he tracked down his former Libby Prison allies and asked them to help him rebut the charges. Their testimony appeared in Sanderson's *My Record in Rebeldom*, dedicated "To My Enemies and My Luke-Warm Friends."

A month after the war ended, the court of inquiry exonerated Sanderson. "The conduct of Col. Sanderson while a prisoner in Richmond deserves praise and not censure," the court concluded. Sanderson was reinstated in the army without loss of pay, but not without lasting damage to his reputation.

On February 11, just four days after General Isaac Wistar's raiders were turned back by Rebel troops at Bottom's Bridge, President Lincoln was contemplating a new raid on Richmond to liberate the captives. The president had learned earlier in the day of the Libby Prison breakout, and he expected the escapees' tales to incite the public to demand action. But that was only one of the reasons that Lincoln was weighing a lightning attack on Richmond. Another was Elizabeth Van Lew's report that the Union prisoners were going to be transferred to the Deep South, where their rescue would be impossible. On the night that he was informed of the Libby breakout, Lincoln sent a telegram to the Army of the Potomac: "Unless there be strong reason to the contrary, please send Gen. Kilpatrick to us here, for two or three days." Twenty-eight-year-old General Hugh Judson Kilpatrick, a bold, cocky cavalry commander and reputed ladies' man, had been promoting a plan to descend upon Richmond and free the Richmond prisoners. Lincoln was ready to give Kilpatrick a hearing.

* * *

Kilpatrick wanted to lead a swift mounted force from the Army of the Potomac's winter encampment to Richmond, seventy miles to the south. General Alfred Pleasanton, commander of the Army of the Potomac's cavalry, doubted that the plan would succeed; a similar raid in 1863 had failed spectacularly. But Pleasanton's superior, General George Meade, optimistically believed that "Richmond might be carried by a *coup de main*, and our prisoners released." Meade ordered Kilpatrick to proceed.

Defending Richmond were just three thousand Confederate militia and cavalrymen and some field artillery. General Wade Hampton's cavalry division, which was nearby, could be expected to quickly come to the capital's aid. But Meade and Kilpatrick did not believe that either General Robert E. Lee's army, facing Meade on the Rapidan River, or other Rebel units in the vicinity could reach Richmond in time to interdict fast-moving Union raiders. Kilpatrick might succeed if they could be kept away from Richmond. Toward that end, General George Armstrong Custer, the "boy wonder" cavalry commander, would lead a sixteen-hundred-man diversionary force against Charlottesville—paralyzing the Rebel units to the west. At the same time, the Army of the Potomac would make two feints in northern Virginia to distract Lee's army.

"Kill Kilpatrick," so nicknamed because of his reputation for squandering men, planned to cross the Rapidan with four thousand cavalry and six guns. At Spotsylvania Courthouse, Colonel Ulric Dahlgren and five hundred troopers would leave the main force and ride southwestward, while Kilpatrick and his remaining thirty-five hundred men proceeded south to Richmond.

At Lincoln's request, Kilpatrick's men would scatter leaflets announcing the president's "amnesty proclamation." A gambit to undermine support for the Confederacy, the amnesty was being extended to Rebel soldiers and civilians who pledged loyalty to the Union (most Confederate leaders were ineligible). Secessionist states where 10 percent of the voters recited the loyalty oath might return to the Union.

Meanwhile, Dahlgren would ride to the James River west of Richmond and cross it there, turn east, and enter the riverside industrial district on the city's south side. Kilpatrick would simultaneously attack northern Richmond to draw defenders away from Dahlgren's men so that they could burn the Tredegar Iron Works, other heavy manufacturers, and the Confederate arsenal, and then free the Union war prisoners. Reuniting north of Richmond, Kilpatrick, Dahlgren, and the liberated war prisoners would march north to join Meade's army on the Rapidan. Success depended on speed, boldness, flexibility, and luck.

A lot was expected of twenty-one-year-old Ulric Dahlgren, who was still recovering from the amputation of his right leg a few days after Gettysburg. He was the son of Rear Admiral John A. Dahlgren, who commanded the Union fleet off Charleston and whom posterity would honor as the "father of American naval ordnance." But Ulric Dahlgren was more than an admiral's son; he had distinguished himself as a cavalryman at Fredericksburg and Chancellorsville before he was wounded in Hagerstown while attacking Rebel cavalry retreating from Gettysburg. When he learned about Kilpatrick's proposed raid, Dahlgren rushed to volunteer, and became the flamboyant general's second-in-command. In a letter to his father, Dahlgren described the upcoming mission with the cheerful fatalism that distinguished his generation: "If successful, it will be the grandest thing on record; and if it fails, many of us will '*go up.*' . . . If we do not return, there is no better place to '*give up the ghost.*'"

✷ ✷ ✷

On the night of February 28, Kilpatrick and Dahlgren forded the Rapidan River, bagging a handful of Confederate pickets, and then rode to

Spotsylvania Courthouse, where they split up. While Kilpatrick's men destroyed telegraph lines and railroad tracks north of Richmond, Dahlgren's troopers galloped over the bone-littered Fredericksburg battlefield and turned southwest in a black, driving rainstorm. They rode hard for two nights and a day, seldom stopping even to eat. At 11 AM on the second day, March 1, Dahlgren's men reached the north bank of the swollen James River west of Richmond, just as planned.

There, the raiders rendezvoused with a slave boy who was to guide them to a ford at Jude's Ferry. But when they reached the crossing, the James was flowing too high and fast to ford. Believing that the boy had deceitfully led them to an impossible crossing instead of the appointed one, Dahlgren ordered him hanged on the spot, using a rein from his bridle. Now, rather than following the James's southern bank into the city's industrial area, Dahlgren's raiders approached the capital from the north side of the river. Along the way they vandalized a canal, wrecked a sawmill, and stole meat and grain from farms.

* * *

Dahlgren learned from slaves in the area that former Virginia governor Henry A. Wise's daughter and War Secretary James Seddon lived at adjacent plantations a half mile apart west of Richmond. He also was told that Wise, now a Confederate general, was on furlough and visiting his daughter. Wise was disliked in the North because he was governor when John Brown was tried and executed in Virginia for his Harper's Ferry raid in 1859. Dahlgren and his men rode up to the Eastwood manor house and demanded to see the general. But Wise and his son-in-law were on their way to Richmond to warn officials that the raiders were near. While his men burned a barn and stables at a nearby plantation, Dahlgren rode on to the Seddon plantation, Sabot Hill, and pounded on the door.

The war secretary's wife, Sarah Bruce Seddon, opened the door and saw Dahlgren standing before her on his wooden leg, leaning on a crutch. When he introduced himself, she inquired whether he was related to Admiral John Dahlgren. He was the admiral's son, the colonel responded.

Mrs. Seddon invited Dahlgren and his staff into her parlor, while telling him that she had been a "beau" of Dahlgren's father when Mrs. Seddon and Dahlgren's mother were schoolmates in Philadelphia. Disarmed by her unexpected connection to his parents, Dahlgren promised Mrs. Seddon that his men would not harm her property. She showed her gratitude by sending a servant to the cellar for a bottle of twenty-year-old blackberry wine.

In the parlor, Dahlgren and Mrs. Seddon drank a glass together while talking about the prewar days when their families were not enemies. While they conversed, some of Dahlgren's restless men burned a Seddon barn, despite their colonel's pledge to his father's former beau, while other troopers vainly searched for a place to ford the engorged river.

The distant sound of gunfire from Richmond's north side ended the raid's strange hiatus. Dahlgren bade Mrs. Seddon farewell and saddled up as snow and sleet fell. Believing that the sound heralded Kilpatrick's anticipated attack on the city, Dahlgren led his force away from the James and rode toward the gunfire, abandoning the plan to free the captives. But when the shooting abruptly stopped, the raiders again changed their plan and rode for Hungary Station north of Richmond, Kilpatrick's and Dahlgren's prearranged rendezvous place. As the snow and sleet reduced visibility to a few yards, the raiders fumbled in the maze of roads and bridle paths outside Richmond.

✷ ✷ ✷

Kilpatrick's resolve began to melt away with the first ragged salvo from Richmond's artillery militia. Rather than attack immediately, he hesitated, while trying to assess the defenders' strength. But the heavy snowfall hid the Rebels like a curtain. As Kilpatrick wavered, the six hundred men in the capital's fortifications better organized themselves and received reinforcements. When the Confederate cannon fire persisted, "Kill" Kilpatrick decided that he had had enough, and broke off the attack. Wet, cold, and exhausted after two days in the saddle, Kilpatrick issued new orders to his thirty-five hundred men. They would abandon the Hungary Station rendezvous site and ride down the peninsula to General Benjamin Butler's lines at Williamsburg. At nightfall on March 1, Kilpatrick's men made camp near Atlee Station, northeast of Richmond.

As they were warming themselves over campfires, Kilpatrick's men were fired upon. Two of General Wade Hampton's Iron Scouts had seen the Union raiders cross the Rapidan and alerted Hampton, who had sent 350 men after Kilpatrick. The Rebel cavalrymen, despite being outnumbered ten to one, charged and chased Kilpatrick's men until dawn. Finally, Kilpatrick managed to halt his raiders' headlong flight and repulse the small Rebel force.

* * *

The alarm bell rang in Capitol Square as the sound of cannon fire from Richmond's northern precincts rumbled through the city. The home defense militia—three hundred factory workers, old men, boys, businessmen, and government workers—dashed to their muster points. Yet instead of dissolving into pandemonium as it had during Wistar's raid, Richmond was preternaturally calm as Kilpatrick probed the city's fortifications. "So suddenly and unexpectedly did this adventure occur, that we were scarcely aware of our danger until it was over," wrote Sally Putnam.

But General John Winder, the Richmond prisons superintendent, was worried. So many prison guards had left their posts to aid the city militiamen that Winder feared a prisoner uprising. Indeed, his concern was justified, for the Libby Prison captives had prepared for just this contingency by organizing themselves into regiments and selecting commanders. When reports of the raid reached them, the prisoners got ready to assist their liberators.

Winder, desperate to prevent Libby's officers from escaping, proposed to mine Libby with explosives. If Richmond's defenses failed, he would annihilate the Union officers before they could escape and lead a citywide insurrection of Union captives.

War Secretary Seddon refused to sanction Winder's proposal. It would have made him a potential accessory to mass murder. But he left Winder a loophole. Under no circumstances, Seddon emphasized, must the captives be permitted to escape.

Choosing to interpret Seddon's second instruction as tacit approval of his plan, Winder ordered several hundred pounds of gunpowder placed in Libby's cellar. If the raiders breached the city's defenses, he intended to detonate the enormous bomb. Major Thomas Turner stood ready to carry out the sanguinary plan, declaring, "I shall stick to my post of duty until Kilpatrick reaches here, then every damn Yankee in this place will be blown to hell."

With picks and shovels, black prison laborers, supervised by Confederate guards, dug a large hole in the cellar and dumped the explosives into it. The Rebels laid powder trains from the cellar room to guard posts outside the building, from where a match might send a flame racing to the gunpowder cache. When the mine was completed, Major Turner bluntly warned the captives that he would blow them "all to Hell" at the first sign of trouble. This failed to impress at least some of the captives. Wrote Lieutenant Cyrus P. Heffley, "This is quite pleasant, to be sleeping on 1,000 pounds of powder. . . . He would not dare put a match to it. Our gov't holds too many of their men."

Kilpatrick's loss of nerve may have spared the prisoners' lives and relieved the Confederacy of later having to justify a massacre of more than eleven hundred men. In a report supporting Winder's radical decision, the Confederate Congress's Joint Select Committee wrote that "indiscriminate pillage, rape, and slaughter" would have occurred if the captives had escaped. "The real object was to save their [the prisoners'] lives as well as those of our citizens. . . . The plan succeeded perfectly. The prisoners were awed and kept quiet." But the raid's failure, not the powder mine, was the reason that there was no uprising. Had Dahlgren's troopers actually reached Libby Prison, some Union officers would undoubtedly have attempted to break out, explosives or not. And it is debatable whether the Rebels, under attack by prisoners and Yankee raiders, would have actually lit the powder train.

✷ ✷ ✷

Dahlgren's five hundred troopers lost their way in the labyrinthian roads and paths around Richmond and became dispersed in the sleet and snow. When Dahlgren at last reached Hungary Station north of Richmond, fewer than one hundred men remained with him; the others had become separated in the woods during a whiteout. Dahlgren looked for Kilpatrick at the rendezvous, not knowing that Kilpatrick had changed his plan and was riding toward General Butler's lines.

A home militia force fired on Dahlgren's party as it crossed the Mattaponi River south of Dunkirk, and Lieutenant James Pollard and twenty-five men from the 9th Virginia rode ahead to alert Captain Edward Fox of the 5th Virginia. Together, they laid an ambush near King and Queen Courthouse.

About 11 PM, after stopping to rest, Dahlgren and the seventy troopers that now remained with him mounted and rode on. Minutes later, they reached a tree-lined stretch of road. As they entered the lane, seventy Rebels who were lying in wait suddenly opened fire. When the shooting stopped, Dahlgren lay dead in the road. Soon, most of his men were prisoners.

✷ ✷ ✷

While ransacking Dahlgren's body for valuables, one of the ambushers, thirteen-year-old William Littlepage, found Dahlgren's notebooks, letters, and other loose papers. They appeared to be important, and so Littlepage turned them over to his teacher and fellow Home Guard member, Edward Hallbach. Littlepage's teacher was stunned by what he read, and without delay forwarded the documents to Richmond. Rebel officials, too, quickly grasped the documents' importance, as well as their extraordinary propaganda value, and released them to the Richmond newspapers. Their publication ignited a furor.

"We will cross the James River into Richmond, destroying the bridges after us and exhorting the released prisoners to destroy and burn the hateful city; and do not allow the rebel leader Davis and his traitorous crew to escape," read Dahlgren's address to his men, written on the stationery of "Headquarters, Third Division, Cavalry Corps." On a similar sheet of paper, under the heading, "Special Orders and Instructions," was written: "The men [prisoners] must keep together and well in hand, and once in the city it must be destroyed and Jeff. Davis and Cabinet killed."

In the explosion of Southern newspaper articles that followed, Dahlgren was given the nickname "Ulric the Hun," and his alleged plan was described as "war under the Black Flag." The *Richmond Daily Whig* asked, "Are they not barbarians redolent with more hellish purposes than were the Goth, the Hun or the Saracen?"

The Union command denied having issued the orders, and claimed they were concocted by the Confederates to fan the flames of Southern hatred against the Yankees. In attempting to discredit the documents, Union officials pointed out that in the lithographic copy sent by General Robert E. Lee to General George Meade, Dahlgren's name was spelled "Dalhgren"; whoever forged the documents, they said, misspelled his name. Yet the instructions appear to be genuine, although Dahlgren probably wrote them without his superiors' authorization or knowledge. Historian James O. Hall believed that Lee's lithographer transposed the letters of Dahlgren's name while touching up the image prior to sending it to Meade. Historian Virgil Carrington Jones speculated that it was a stenographer at Kilpatrick's headquarters who committed the error. Jones concluded that Dahlgren wrote the instructions, because Dahlgren's private memo book contained a nearly verbatim pencil draft.

Genuine or counterfeit, the instructions outraged Confederate civilians and soldiers, and the Rebel command began planning a retaliatory raid against Ohio and Illinois. First lady Varina Howell Davis professed bewilderment in her journal: "Once Commodore Dahlgren [Ulric's father] had brought the little fair-haired boy to show me how pretty he looked in his black velvet suit and Vandyke collar, and I could not reconcile the two Ulrics."

* * *

After rifling through Dahlgren's possessions, a soldier severed the fifth finger of the colonel's right hand and pocketed the gold ring that Dahlgren wore in memory of his dead sister. The Rebels stripped the corpse and dumped it in a muddy hole at a crossroads. Later, when Confederate officials demanded that it be sent to Richmond, the militiamen dutifully exhumed the body, dressed it in a shirt and trousers, lay it in a wooden coffin, and transported it to the capital.

In Richmond, the coffin was opened for public viewing at the Richmond & York River Railroad Station, and Dahlgren's wooden leg was displayed in a nearby shop window. "His face wore an expression of agony," reported the *Richmond Daily Whig*. Scores of people came to gawk at the ghoulish exhibits—until Jefferson Davis ended the carnival by requiring that the remains be buried in secrecy. "It was a dog's burial, without coffin, winding sheet or service," reported the *Daily Richmond Examiner*.

Now it was the Union's turn to be offended. In a letter on March 11, General Benjamin Butler requested that the desecrated body be handed over so that it could be sent to Dahlgren's admiral father. "Some circumstances of indignity and outrage accompanied his death. You do not war upon the dead," Butler lectured Colonel Robert Ould. Surrendering Dahlgren's body would partly mitigate the Union's anger over its insulting treatment, the general wrote. Surprisingly, the Confederates consented. But when gravediggers reopened Dahlgren's resting place in Oakwood Cemetery on April 14, the grave was empty. "Dahlgren had risen, or been resurrected," the *Examiner* wryly observed.

* * *

Elizabeth Van Lew had been appalled by the press reports of March 7 and 8 about Ulric Dahlgren's death and the abuse of his corpse. She asked her fellow Unionists to locate the burial site. It wasn't easy, but the Unionists at last found a black cemetery worker who, from a hiding place behind a tree, had witnessed the furtive interment.

On the night of April 6, the cemetery worker led Unionists to the burial spot. They exhumed the body and moved it to an outbuilding on Unionist William Rowley's property. Van Lew, her mother Eliza, and other members of Richmond's underground paid their respects to the dead raider. They examined the remains with "gentle hands and tearful eyes," Van Lew reported. Except for the head, she wrote, the body "was in a perfect state of preservation, fair, fine and firm the flesh. . . ." The Unionists cut off locks of Dahlgren's hair to send to his father before laying him in a metal coffin.

Rowley concealed the coffin beneath peach tree saplings in his wagon bed, and then drove the wagon past a talkative Confederate guard at a roadblock without being searched. Reaching the farm of a Union sympathizer outside Richmond, he reburied the coffin under one of the saplings. The Richmond underground informed Admiral Dahlgren—after swearing him to secrecy—but no one else knew the details of their grave robbery until after the war. Just as the Libby Prison breakout became known in Richmond as "the great escape," the purloining of Dahlgren's body became "the great resurrection."

* * *

Judged a failure by Union and Confederate observers alike, the Kilpatrick raid cost the lives of six Union soldiers. Twelve others were wounded, and 317 were reported missing—most of them captured and sent to Richmond war prisons, where they were singled out for special punishment. Meanwhile, General Braxton Bragg and War Secretary Seddon proposed public executions for some of the captured raiders. But General Robert E. Lee argued that many of Dahlgren's men were probably ignorant of Dahlgren's incendiary instructions, which might not have even been sanctioned by the Union command. "I do not think it right, therefore, to visit upon the captives the guilt of his [Dahlgren's] intentions," wrote Lee. He also warned that executions would invite Union retaliation against Confederate captives. None of the raiders was put to death.

Eight officers and a surgeon from the Kilpatrick raid were marched to Libby Prison to join Colonel Thomas Rose and the recaptured escapees in the cellar. But first Colonel A. C. Litchfield and two fellow officers were required to stand under guard at the prison entrance for three days, so that Richmond civilians could publicly revile them. One of their worst tormentors was the war secretary's wife, Sarah Seddon. The raiders had damaged her property while she sipped wine with Colonel Dahlgren, and "her rage exceeded all previous exhibitions," Litchfield ruefully reported. She vowed to "use her influence to have us put in dungeons, and fed on bread and water till we rotted."

Mrs. Seddon was as good as her word: for four and a half months, the raider officers shared a 12-foot-square cell with four black prisoners. A reeking tub, rarely emptied, served as their privy. They consumed their bread and water under guard, without utensils. Forbidden to light a fire, they lived in perpetual gloom, the boarded windows admitting only gleams of light through the cracks. "We were kept as a sort of menagerie for exhibition to the curious negro-breeders and negro-haters, all delighted that the Yankees had found so fit companions," wrote Litchfield.

* * *

Although the *Richmond Dispatch* shrugged off Kilpatrick's raid as nothing more than "robbing hen-roosts and stealing negroes," Richmond had had a close call. "My observations convinced me that the enemy could have taken Richmond," General Wade Hampton conceded in his report. Seddon believed that Richmond was spared only because the Union attack was "conducted with a timidity and feebleness that were in ludicrous contrast with the boldness of the conception and the extent of the means"—an accurate assessment of Kilpatrick's loss of nerve. The raid reminded Richmond residents of their vulnerability. "It is humiliating indeed that he [the enemy] can come and threaten us at our very gates whenever he so pleases," wrote Mary Chesnut.

* * *

Hastily convened in response to the Libby escapees' harrowing accounts of their captivity, the Joint Select Committee on the Conduct of the War interviewed escapees and newly paroled convalescents in Washington, Annapolis, and Baltimore. In May 1864, it issued a thirty-page report, whose findings War Secretary Edwin Stanton summarized: "There appears to have been a deliberate system of savage and barbarous treatment and starvation." The report, accompanied by eight sketches of shockingly emaciated new arrivals at Annapolis, condemned the Confederate prison system as a cruel instrument whose purpose was to break down and destroy Union captives. During and following the report's compilation, some of the crippled survivors, unfit to return to duty, hobbled through the Capitol corridors on crutches, lobbying on behalf of their still-imprisoned comrades. Partly due to their efforts, Congress eventually would launch a lengthy, far-reaching investigation of Southern war prisons, focusing largely on Andersonville and its unique horrors. The report filled twelve hundred pages.

Three months later, the United States Sanitary Commission released its own report on Confederate prisons, which included more than two hundred pages of testimony and observations as well as photographs of cadaverous Union parolees. "Tens of thousands of helpless men have been and are now being disabled and destroyed by a process as certain as poison, and as cruel as the torture of burning at the stake." By contrast, the report said, the Union treated Rebel prisoners with "all the consideration and kindness that might be expected of a humane and Christian people."

* * *

In the early spring of 1864, thousands of people wearing their Sunday best waited beneath the blooming dogwoods in Richmond's Capitol Square to greet the first Confederate prisoners to return since 1863. The welcome news of the Confederate parolees' homecoming was announced earlier in the day from Richmond's church pulpits. As the soldiers marched into the square to the lively strains of "Dixie," cheers rent the scented air. Jefferson Davis made a speech, and the crowd shouted, "Hurrah for the graybacks!" The people pressed bouquets and delicacies on the soldiers, some of them looking better fed than the civilians making gifts of their hoarded food.

Prisoner exchanges had resumed after intensive negotiations by General Benjamin Butler and Colonel Robert Ould, for once united in purpose, although inspired by divergent motives. While Butler wished to rescue Union captives from life-jeopardizing conditions, Ould sought fresh Rebel troops for the armies of Lee and Joseph Johnston, and to rid the South of the burden of caring for the Union captives. For the greater good of freeing the Yankee prisoners, Butler chose to overlook the Confederacy's refusal to exchange black prisoners and its violation of thousands of Rebel paroles that were granted when Vicksburg surrendered. Thousands of man-for-man exchanges were made in late March and early April.

The exchanges lasted until General Ulysses Grant ordered them stopped during his first trip to Fortress Monroe as general-in-chief. "We got no men fit to go into our army, and every soldier we gave the Confederates went immediately into theirs," Grant complained to Butler. The Union held 26,000 rested, decently fed and clothed prisoners, wrote Grant, and if exchanged, they would comprise "a corps, larger than any in Lee's army, of disciplined veterans."

Abraham Lincoln's previous army commanders had been incompetent or timid, but Grant was neither. Lincoln believed Grant was the one who would at last crush the Confederacy. Grant's blunt strategy was to "concentrate all the forces possible against the Confederate armies in the field" at several points simultaneously—thereby preventing them from reinforcing one another—and to then destroy them. At the same time, he intended to systematically deny the enemy every means of continuing the war. That meant ending prisoner exchanges.

Butler agreed with Grant, but warned him that there would be a public outcry if the exchanges were stopped "simply on the ground that our soldiers were more useful to us in rebel prisons than they would be in our lines." However, if the stated reason was the South's refusal to release black prisoners, the Northern people would support the suspension, Butler said.

Grant accepted Butler's advice. On April 17 he formally notified Butler that before exchanges could resume, the Rebels must meet two conditions: exchange black Union captives with "no distinctions," and balance out the improper return to the ranks of the Vicksburg parolees by releasing an equivalent number of Yankees.

By August the Northern prison camps were swollen with 67,500 Rebels, while 50,000 Union captives languished in the Confederacy's prisons. Butler proposed a meeting with Ould to discuss resuming man-for-man exchanges, but Grant reminded him of his reasons for wishing to continue the cartel's suspension. "It is hard on our men held in Southern prisons not to exchange them, but it is humanity to those left in the ranks to fight our battles," Grant wrote. "Every man we hold, when released on parole or otherwise, becomes an active soldier against us . . . to release all rebel prisoners in the North would insure Sherman's defeat and would compromise our safety here [in Virginia]." The general-in-chief made no apologies for his flinty policy, either, when he testified about the cartel's suspension before Congress in 1865.

Yet, except for the Union command and some Lincoln administration officials, nearly everyone in the North believed that the Confederacy's willful refusal to release the black prisoners was the reason the exchanges had stopped. But there were skeptics. Walt Whitman wrote in a letter to the *New York Times*, "In my opinion, the anguish and death of these ten to fifteen thousand American young men, with all the added and incalculable sorrow, long drawn out, amid families at home, rests mainly upon the heads of members of our own government."

When Butler entered politics after the war, serving in Congress and as Massachusetts's governor, he stated explicitly that Grant, and not he, had stopped the exchanges. In his autobiography, Butler wrote that the Union prisoners who died in Southern captivity gave their lives "as a part of the system of attack on the Rebellion, *devised by the wisdom of the general-in-chief of the armies* [author's emphasis] to destroy it by depletion, depending upon our superior number to win the victory at last."

* * *

As Grant's spring offensive ground toward Richmond, the Confederacy, just as Elizabeth Van Lew had warned, began transferring its Richmond war captives to Andersonville and other prisons in the Deep South. "We take the cars and are on our way for—we don't know where. Further South," Lieutenant Cyrus P. Heffley scribbled in his diary after he and his Libby Prison mates were awakened at midnight on May 7 and were told to be ready to march in one hour. Before leaving, the officers destroyed the food they could not take with them, as well as their homemade furniture, and they tried to burn Libby down. Fifteen hundred officers were squeezed into boxcars and transported to Macon. Some later went on to Charleston.

Belle Isle's 10,000 cadaverous, ragged prisoners at last departed the cruelly misnamed island. Conditions had worsened since winter. The men lived in pits clawed in the ground and a dozen a day died; starving prisoners trapped, cooked, and consumed dogs belonging to Rebel officers; and men driven mad by hunger sometimes ate their comrades' vomited breakfasts. After the Rebels paroled those deemed too ill to survive a 500-mile journey in a boxcar, the enlisted men— gaunt, sunken-eyed, filthy, some of them deranged—were marched over the footbridges to entrain for their new camp, not imagining they were going to a worse place. But they were. Andersonville, the apotheosis of suffering, would one day be a byword for living death.

Although it had opened only in March 1864, by summertime Andersonville was so densely populated—30,000 men occupied a stockade designed to hold 10,000—that prisoners often fought for a tiny piece of ground of their own. A filthy stream provided water for drinking and bathing, and—in the case of the Rebels camped upstream— served as a privy. Before drinking the dirty water, the prisoners filtered it through their shirts. Scurvy turned their gums black, their teeth fell out, and some of them lost their hair and fingernails. Dozens of Union captives died each day of dysentery, malaria, and diarrhea, with some of the victims weighing as little as 60 pounds when they expired. (Later, when conditions were at their worst, up to a hundred prisoners died each day.) They were buried without coffins, or even minimal dignity, as one captive reported, "their hands in many instances being first mutilated with an ax in the removal of any finger rings they may have."

From March through May, 1,567 Andersonville prisoners died, and between May and August, 6,000. Without warning, the guards shot and killed them for trying to escape, for crossing the "dead line," and sometimes, for merely sitting quietly in front of a fire. Disobedient captives were punished in the "standing stocks" and shackled to 32-pound cannonballs.

After his August 1864 tour of Andersonville, Confederate inspector general D. T. Chandler arraigned the "criminal indifference of the authorities" and entreated his superiors to send no more prisoners. The prison was, Chandler wrote, "a reproach to us as a nation." Indeed, when six hundred Andersonville prisoners were released in December 1864, the extent of their suffering became widely known. "In many instances the bones of hips, spine, and shoulders projected bare through the skin," wrote a horrified Lieutenant Asa Isham, a one-time Libby Prison prisoner who accompanied the parolees from Charleston, his comparatively humane recent place of captivity.

In June, General John Winder, the Richmond prisons commandant, was placed in charge of the prisons at Macon and Andersonville.

After inspecting Andersonville, Winder requested more guard units, tents, and surgeons. He punished rogue guards who abused and killed prisoners without justification. He also planned to transfer many Andersonville captives to a new prison in Alabama to relieve the stockade's dreadful overcrowding. But Winder received little of what he asked for. The Confederacy was grappling with the twin crises of Grant's Overland offensive in Virginia and General William Sherman's campaign in Georgia—the latter also dooming Winder's proposed Alabama prison.

Winder had long lobbied for a central command to replace the discordant management of the Confederate war prisons. In late 1864, he got his wish: he was appointed commissary-general of all the Rebel prisons east of the Mississippi River. But the Confederacy had waited too long. Winder presided over a broken system before succumbing to a heart attack in early 1865. Had he survived the war, he might have been hanged with Captain Henry Wirz, the notorious Andersonville warden and former Libby commandant. As it was, Winder was posthumously indicted as a co-conspirator with Wirz, Jefferson Davis, and other Confederate leaders. Wirz, however, was the only one executed for war crimes—namely, for the mass murder of nearly 13,000 Yankees in less than one year, and for maiming and disabling many thousands more.

* * *

With each passing month, the Confederacy's ability to feed and care for its growing captive population slipped further. In desperation, Rebel officials encouraged the Andersonville prisoners to petition the Union government for their release, while advancing an array of ideas for restarting the exchanges and sparing captives' lives. They proposed that each side send surgeons to distribute food and clothing and provide medical care to its imprisoned troops. The Confederacy urged the Union to sell it medicine for the war captives, and to resume unlimited man-for-man exchanges. The Rebels also offered to release thousands of seriously ill Yankee prisoners, if the Union would only transport them home. The Union command ignored all of the overtures except the last, and four months passed before it supplied transportation for the broken-down parolees.

As Libby's population waxed and waned for the last time, the Confederate Congress issued a new war prisons report that resembled its earlier reports. It denied that Union prisoners were mistreated, or robbed, or that they received skimpier rations than the Confederate soldiers who guarded them. The captives were treated with "humanity and kindness," the report concluded. Grant characterized the report's assertions as "an infamous lie."

In January 1865, the dying Confederacy finally consented to exchange black Union prisoners, but the concession was made too late to benefit the South militarily or logistically. Its armies shattered, in retreat, or clinging to hopeless defensive positions, the Confederacy no longer could reasonably hope to delay defeat much longer. The Union command noted that recently captured Rebel soldiers were often in deplorable physical condition and barefoot, toting haversacks that were empty save for a pint of corn kernels and three inches of bacon no thicker than a pencil—their ration for three days. With all the signs portending a Union victory, Grant saw no harm in resuming exchanges, and thousands of captives went home. Three months later, the war ended.

* * *

Figures released immediately after the war suggested that for all the Union's protestations of Rebel mistreatment of Yankee captives, the same percentage of Confederates died in Union war prisons as Yankees in Southern prisons—a little more than 11 percent. The surprising numbers showed that 24,000 of the 215,000 imprisoned Confederates had died in Union captivity, compared with 22,000 Yankee deaths among the 194,000 held in Rebel prisons. Southern historians cited these figures for a century or more as evidence that the Yankee captives fared no worse than the Rebel prisoners in the North.

Today it is clear that the death totals were incomplete and inaccurate. While every Union prison kept a record of prisoner deaths, many Rebel prisons kept no record at all. Historians and scholars have since revised the Yankee death toll upward to more than 30,000—at least a 14 percent mortality rate—based on unit records and grave sites at 16 former Confederate prisons.

In 1876, eleven years after the war, the question of whether Union captives had been mistreated remained very much alive. "It is admitted that the prisoners in our hands were not as well provided for as we would [have wished], but it is claimed that we did as well for them as we could. Can the other side say as much?" Jefferson Davis wrote that January. Later that year, in *Confederate View of the Treatment of Prisoners*, Southern leaders accused the Union of having spread "the most malignant stories . . . [of] 'Rebel barbarities'" to conceal the Union's momentous decision not to resume exchanges. "The real cause of the suffering on both sides was the stoppage of the exchange of prisoners, and for this the *Federal authorities alone* were responsible," said *Confederate View*.

In fact, this was essentially correct. Many Union prisoners who died in 1864 at Andersonville and elsewhere might have survived had not the Union stopped the exchanges in order to wage "total war." While the Confederacy strove to restart the cartel, the Union command cynically used the Rebels' refusal to exchange black prisoners and the parole violations after Vicksburg as excuses not to. Meanwhile, Rebel prisoners in the North were being deprived of food in retaliation for the South's growing incapacity to care for its Union captives, whose numbers soared after the cartel's suspension. It is almost certain that if the Confederates had capitulated in 1864 on the sticking points cited by the Union, another reason would have been found to not restart the exchange. In the war's fourth year, Lincoln and Grant were willing to adopt any measure, no matter how harsh, if it more quickly ended the suffering and dying. In retrospect, they were probably right. While thousands of Union captives perished when prisoner exchanges stopped, many more than that would have died in battle had the Union resumed exchanges and released 26,000 combat troops to the Confederacy, prolonging the bloodletting.

The Civil War, and especially the Union's total war doctrine of 1864 and 1865, obliterated the last vestiges of the eighteenth-century European mode of "civilized" warfare. In no major war fought since then has there been a prisoner exchange cartel.

EPILOGUE

*"When a people all think and speak one way, there is not much
liberty left among them. Here the people are a unit."*

–Elizabeth Van Lew, October 27, 1876

APRIL 3, 1865, RICHMOND

On the day that Richmond had dreaded for so long, just one person
remained in Libby Prison—Major Thomas Turner. The Army of the
Potomac was marching unopposed on the Rebel capital, and it would
soon be in Yankee hands. After the Confederate government had en-
trained to Danville with the government's records and gold, and the
Rebel defenders had marched away, Mayor Joseph Mayo formally
surrendered his city. Libby Prison was empty, save for Major Turner,
who was gathering up his belongings and burning records. When he
had completed these tasks, he stepped into the street. The streets,
too, were full of smoke; Richmond was burning. Turner looked for a
way out.

As the Rebels were abandoning the capital, they set fire to the waterfront warehouses, which held millions of dollars' worth of tobacco,
and to the city's celebrated flour mills. The conflagration quickly
spread to adjoining buildings and then raged through the city's central
business and residential districts. At the same time, Rebel naval officers detonated the magazines of the Confederate Navy's anchored gun-
boats. The bone-rattling explosions broke windows two miles away.
Flames leaped from three James River bridges that had been set ablaze.

A powder magazine was blown up on the city's north side, killing eleven people in a nearby almshouse. More than three hundred convicts escaped from the state penitentiary when the guards fled. Looting broke out. The authorities poured barrels of whiskey into the gutters in the hope of averting wholesale drunkenness, but the looters dipped cups and pans into the streams; some lapped up the whiskey like dogs. Elizabeth Van Lew described the wild scenes as "the consummation of the wrongs of years."

Van Lew retrieved the large American flag that had been smuggled to her from General Benjamin Butler and ran it up her flagpole, hoping Union troops would see it when they marched into Richmond from the east. Hers was the first Union flag to fly over the city in four years.

The victors in blue entered the enemy citadel without bloodshed. As the Yankees marched down Main Street, black men and women packed the sidewalks—praying, shouting, and darting into the street to embrace the soldiers and their horses. A Union band played "Dixie" as it strode through the smoke. The conquering army organized fire-fighting brigades and put out the fires, but not before nine hundred buildings were gutted in the heart of the city.

*   *   *

Confederate soldiers visited the Van Lew mansion with the intention of burning it down. Van Lew's niece, Annie, described how "Auntie went out on the front porch and said if this house goes, every house in the neighborhood should follow." Union troops, Van Lew warned, would take vengeance against the homes of Confederate loyalists. The Rebels evidently believed her, for they left without causing any damage.

Unaware that the Rebels had already been there, General Ulysses Grant sent Lieutenant David Parker to the Van Lew home to protect one of his most valuable spies. Parker, who supervised the mail service for the Army of the Potomac, rode up to the big house on East Grace

Street after first securing Richmond's post office. Van Lew declined Parker's offer to post a guard around her home, but she invited him to dinner. The dining room was filled with Unionists who had served in the Confederate bureaucracy and had passed intelligence to Van Lew, and thence to the Union command. One of Parker's tablemates made a particular impression—Erasmus Ross, the Libby Prison clerk. Van Lew told Parker that she had gotten Ross hired at Libby and that he had been "doing my bidding" since then.

Two days later, President Abraham Lincoln and his son Tad, with a bodyguard of ten sailors, toured the devastated city. Everywhere the president's party went, a crowd of black servants and former slaves followed. When they passed Libby Prison, someone shouted, "We'll pull it down!" Lincoln shook his head. "No, leave it as a monument," he said.

When Ulysses Grant and his wife, Julia, came to Richmond, they called at Elizabeth Van Lew's home and drank tea with her on the veranda. The general-in-chief personally thanked Van Lew for her services to the Union Army. He later wrote her a letter expressing his gratitude: "You have sent me the most valuable information received from Richmond during the war."

After reducing the documents in his Libby Prison office to ashes, Major Thomas Turner slipped out of the burning city. He managed to reach General Robert E. Lee's retreating army in southern Virginia, but when Lee surrendered at Appomattox Courthouse, Turner then set out to join General Joseph Johnston's army. When he reached Augusta, Georgia, however, Turner learned that Johnston had capitulated in North Carolina. He now headed west, crossing the flooded Deep South on foot and by dugout canoe and sleeping in the woods. He rode through Arkansas on a mule. In Texas, he joined Confederate general Jubal Early in traveling to the Bahamas, and thence to Havana. Turner was in Cuba when he learned of the hanging of Henry

Wirz, Andersonville's commandant. Vowing not to become another victim like Wirz, "a poor defenseless foreigner," he sailed to Canada. Ten years passed before Turner returned to the United States. He practiced dentistry in Tennessee and died in 1901.

✶ ✶ ✶

Libby's hated jailkeeper, Richard Turner, did not manage to escape. Immediately after the war, he was cast into Libby's dungeon, where he had condemned so many Yankees to weeks of misery. While the prison held more captives than ever—nearly three thousand Confederate prisoners—they slept on beds and had adequate food. Turner's five-year-old son, William, supplemented his father's ration by kicking parcels of food to him in the cellar through the barred ground-level window. Turner did not remain there for long. With a knife smuggled to him by his wife, he sawed through the wooden bars—which had survived the iron bar retrofitting a year earlier—and escaped, becoming possibly the only prisoner to break out of Libby's cellar jail. After he had been at large for a month, Turner was recaptured while visiting his family in Richmond, and was sent to the more secure state penitentiary. People came from all over to see the infamous "Libby Lion."

The violent feeling against Turner might have resulted in his being executed beside Wirz, had it not been for the intervention of another accused war criminal, Captain Richard Winder, a nephew of the late Richmond prisons commandant. One night, Winder plied the duty officer with brandy and persuaded him to show him the written charges against Turner. As soon as the documents were in his hands, Winder destroyed them—and with them, the Union's case. As a result, Turner's trial was expedited, and he served only one year in prison before his release in June 1866.

When Richard Turner died in 1901—also the year of Major Thomas Turner's death—he was Democratic Party chairman of the southeastern Virginia county where he lived.

* * *

Two months after Rebel troops overpowered Colonel Thomas E. Rose outside Williamsburg and marched him back to Richmond, he was led from Libby Prison's dungeon into the blinding sunlight. His filthy, tattered clothing hung on his shrunken frame. But his day of liberty was now at hand. With thirty-three other paroled Union officers, Rose boarded a flag of truce boat, the *New York*, at City Point. "The Stars and Stripes float over me again for the first time in more than ten months," wrote Captain Robert T. Cornwell, another of the parolees. Most of those aboard, including three hundred seriously ill Yankee enlisted men, would not return to the army, either because of broken health, battle wounds, or other disabilities.

But a month after his release, Colonel Rose rejoined the 77th Pennsylvania. The regiment served with General William Sherman's army in Georgia until Atlanta fell, when it was sent to Tennessee. There, the 77th was in the middle of the fighting at Franklin and Nashville, and it pursued General John Bell Hood's army into Alabama. When the 77th was mustered out in Philadelphia in January 1866, Rose was a brevet brigadier general.

Preferring military life to the classroom, he joined the regular army, where he remained until his retirement in 1894. He died in 1907 in Washington, D.C. His headstone in Arlington National Cemetery cites the many battles in which he fought, while also noting arguably his greatest triumph:

ENGINEERED AND EXECUTED THE LIBBY PRISON TUNNEL

* * *

Neither General Neal Dow nor Colonel Abel Streight saw combat again with the Union Army. Streight mustered out in March 1865 as a brevet brigadier general and returned to Indianapolis, where he served in the state senate and operated his publishing business until his death in 1892. Dow resigned his commission in November 1864. Embittered by his experience as a war prisoner, he lobbied for harsher measures against the post-war South, while continuing to crusade for prohibition. He died in Portland, Maine, at the age of ninety-three.

* * *

Libby Prison reverted after the war to its original purpose, a warehouse; its major tenant was the Southern Fertilizing Company. In the 1880s, seven Chicago businessmen, including sports magnate A. G. Spalding and candy-maker Charles F. Gunther, purchased the warehouse for $23,000. The Chicago men formed the Libby Prison War Museum Corporation and announced an audacious plan. They would move the infamous war prison from Richmond to Chicago and reopen it as a public attraction.

In 1888, the warehouse was painstakingly dismantled, brick by brick, timber by timber, capstones and windowsills. Meticulously numbered and catalogued, the 600,000 pieces were loaded into 132 railroad boxcars and shipped to Chicago—where they were reassembled as the Libby Prison War Museum.

At the grand opening on September 20, 1889, a brass band played old war tunes as hundreds of former officers of the Grand Army of the Republic walked through the prison museum. Some of the approximately thirty-five hundred Libby Prison alumni who visited during the next several years were inspired to form the Libby Prison Tunnel Association, presided over by Harrison Hobart, the fireplace gatekeeper during the great escape.

The museum was such a faithful re-creation of their prison that chills must have run down the former captives' spines when they saw the familiar rooms, and the timbers and windowsills that still bore the carved names and initials of old comrades. Also exhibited were Civil War–era pistols, muskets, carbines, cannons, and torpedoes; a piece of the pillow where Lincoln's head rested when he died; Ulysses Grant's death mask; a roof plate from the ironclad *Merrimack*; the table upon which generals Grant and Lee signed surrender documents; a piece of the flag that flew over Fort Sumter when it was surrendered in 1861; the manuscript of Lee's farewell to the Army of Northern Virginia; and the chisel—the one Major B. B. McDonald had let Elizabeth Van Lew hold—that helped dig the Libby tunnel. And, of course, there was a replica of the famous tunnel entrance.

Not every former inhabitant of Libby Prison was impressed. George Putnam thought it "a stupid plan, for the historic interest of the building was properly to be connected with its location." He also found "something repellent in the thought of using as a showplace a structure which represented so much of pathetic tragedy."

Still, the museum thrived—100,000 people came during the first three months—until the 1893 Chicago World's Fair lured away visitors with its glitzier exhibitions. They never returned in their previous numbers. In just a few years the museum became insolvent, and in 1899 it was razed. Utilizing part of Libby Prison's façade, the Chicago Coliseum rose in the museum's place and became a venue for professional sports and rock concerts. When the Coliseum was torn down in 1982, the Chicago Historical Society added the historic façade to the bricks and war artifacts from the museum already on display in the Society's Civil War Room. Libby's beams and timbers, rich with carved captives' names and unit numbers, were purchased by a farmer from LaPorte, Indiana, who used them to build a barn. A real estate agent later purchased and dismantled the barn, intending to re-erect it elsewhere. He never did.

Descendants of Civil War veterans from the North and the South dedicated a plaque in 1980 at the prison site in Richmond. The marker is the only evidence that Libby Prison once stood there.

* * *

For more than a century after the Civil War, Belle Isle, as before the war, was the site of Old Dominion Iron and Steel Co.'s operations. Later, Virginia Electric and Power Co. built a hydroelectric plant there.

In 1973, the "pretty island" where so much suffering and death had occurred became part of Richmond's James River Park System.

* * *

The war had cost Elizabeth Van Lew not only her neighbors' goodwill, but most of her assets—spent on bribes, food, and supplies for the Union captives, and information for the Union command. Between 1860 and 1870, the Van Lews' personal worth plunged from $37,000 to $15,000. Van Lew hinted to her former Union Army contacts that a government emolument would be welcome, and they lobbied on her behalf. General George Sharpe, the former intelligence chief for the Army of the Potomac, recommended that Congress give her $15,000. "For a long time, she represented all that was left of the power of the U.S. Government in the city of Richmond," Sharpe wrote in a supporting letter. Van Lew traveled to Washington in her quest for financial aid. After meeting with her, General Benjamin Butler wrote, "I retain a lively sense of your patriotism and fidelity to the Country in her darkest hours." But the government did not grant Van Lew's request.

The War Department did agree to return to her all of the wartime messages that she had sent to the Union command—coded, in invisible ink, and otherwise. To protect her wartime spy network from postwar reprisals, she destroyed all of the correspondence except her January 30, 1864, letter warning Butler that the Confederacy planned to transfer the Richmond prisoners to Georgia.

Ulysses Grant never forgot Van Lew's valuable services and sacrifices. Thirteen days after being sworn in as the eighteenth president in 1869, Grant appointed her Richmond postmistress—a position that paid $4,000 annually; it was one of the highest federal offices that a woman could hold. The appointment affronted the Richmond establishment. "We regard the selection of a Federal spy to manage our post-office as a deliberate insult to our people," fumed the Richmond *Daily Enquirer and*

Examiner. Van Lew hired black clerks and mail carriers and spoke openly in favor of civil rights.

Van Lew was often reminded that she was regarded as a traitor to the South. She and her mother were refused admittance to some Southern resorts. Few people called at the Van Lew mansion on Church Hill, and practically no one in Richmond even wished to be seen conversing with them. "When a people all think and speak one way, there is not much liberty left among them. Here the people are a unit," Van Lew wrote.

As postmistress, Elizabeth Van Lew did not last beyond Grant's eight years in office; in 1877, President Rutherford Hayes replaced her. Several years later, financial necessity compelled her to take the civil service examination and move to Washington, D.C., to become a postal clerk, earning $1,200 a year. But she clashed with her supervisor, and she was demoted to the dead letter office, where her salary was reduced to $720. She resigned and returned to Richmond.

There she discovered that the hostility toward her had intensified. Van Lew was shunned by adults and taunted by children. A man stopped on the street to tell her, "We can never forget it nor forgive you." The city was in the throes of mythologizing the "Lost Cause"; a culminating event was the unveiling of a statue of Robert E. Lee before 100,000 people. It all served to remind Richmond residents that Van Lew had once been a secret enemy of their beloved Confederacy. She tried to sell her Church Hill mansion so that she could move to the North, but she did not receive a single offer.

Lacking funds to leave Richmond, Van Lew became a virtual prisoner in her home. "[I am] held in contempt & scorn by the narrow minded men and women of my city for my loyalty," she wrote. "No one will walk with us on the street. No one will go with us anywhere. . . . We are held so utterly as outcasts here. . . . It is an awful way to live."

In an article published after Van Lew's death, the *Richmond Evening Journal* described her during these years as a "small twisted figure," whose angular face was framed by shoulder-length white curls. "She seemed a witch of a woman—a strange, uncanny creature, muttering and talking to herself as she walked the streets." Her "shadowy garden, and the silent house with its solitary taper, was a spooky place at night."

Elizabeth Van Lew died in 1900 at the age of eighty-one and was buried beside her parents in Shockoe Cemetery.

* * *

During the war, the observant Annie Van Lew noticed that her aunt kept the code key for her ciphered correspondence in her watchcase. By consulting this chart of seemingly random numbers and letters, Van Lew was able to encrypt and decrypt her correspondence with the Union command and the Richmond underground. After her death, as Van Lew's nieces and other relatives sifted through her belongings in the dusty mansion, they found the old watch. Remembering the code key, they opened the watchcase, and there, in the place where it had lain for nearly forty years, was the yellowed piece of paper.

A few months later, a Bostonian who was attending a bankers' convention in Richmond toured the old Church Hill mansion, and asked the guide to lead him to the attic. Beneath the south eaves, the Yankee banker pointed to an 18-inch-wide wall panel and told the guide, "Here I passed fourteen days during the war." The hidden room ran the length of the house and, according to the old soldier, could conceal at least fifty men.

In 1901, the Virginia Club purchased the mansion for $13,000, installed electric lights, and turned it into a clubhouse. The club's servants soon were reporting strange noises and ghostly apparitions, some of them supposedly bearing an eerie resemblance to Elizabeth Van Lew.

The Virginia Club moved to another location, and the old house became a sanitarium. In 1911 it was condemned by the city and torn down. Bellevue Elementary School today stands on the Van Lew mansion site.

* * *

In 1902, Elizabeth Van Lew's Massachusetts friends brought a granite boulder from Boston's Capitol Hill to Richmond, and placed it over her grave. Set in the rough-hewn rock is a bronze tablet bearing this inscription:

> *She risked everything that is dear to man—friends,*
> *fortune, comfort, health, life itself, all for the*
> *one absorbing desire of her heart—that slavery*
> *might be abolished and the Union preserved.*

ACKNOWLEDGMENTS

Several libraries and their staffs assisted my research immensely. Without their efforts, this book would have been impossible to write.

Chief among them is Duke University's Special Collections Library. The gracious, experienced librarians there displayed genuine interest in my project and, for weeks on end, cheerfully retrieved family papers, diaries, documents, photographs, and other materials from storage. They were never impatient; they invariably volunteered to do more.

In Richmond, the staff at the Virginia Historical Society library went the extra mile to make sure that I found everything I needed during my days of research there. It is hard to imagine a more pleasant working environment than the Virginia Historical Society library.

The Library of Virginia was another excellent resource when I was doing research in Richmond, and its staff came to my aid when I needed help.

At the University of North Carolina in Chapel Hill, the Walter Royal Davis Library once again served as my "home" library. The reference desk staff tolerated my innumerable comings and goings from the basement, which is the repository for a large collection of congressional records. Also of assistance to me at UNC were the staffs at the Louis Round Wilson Library and its rare books and North Carolina collections, and at the Park Library of Journalism and Mass Communication.

Donna Humphrey of the Bucks County, Pennsylvania, Historical Society has my heartfelt thanks for helping to track down the elusive historical material on Colonel Thomas Rose. She dug through the archives in Rose's home county and helped me flesh out the background on a man who in other accounts appeared to spring to life full-grown in the garb of a Union Army officer.

Paul Hogoian at the Library of Congress's photoduplication service stepped up when he learned there had been a delay in getting illustrations into my hands, and e-mailed them to me that day. It is reassuring to know that there are government employees like Paul who are conscientious, competent, and helpful.

Cornell University deserves credit for putting online all of the official records, correspondence, and reports of the Union and Confederate armies and their high commands. It is a privilege to be able to read these historical documents without leaving one's study.

Lindsay Jones at PublicAffairs helped make this a better book than it otherwise would have been. I am grateful for her suggestions and cheerful support.

Finally, without the support of my wife Pat, research scientist and history buff, this book and its predecessors would not have been possible.

NOTES

PROLOGUE

ix **After enduring several weeks:** Tobie, pp. 43–46.

x **Caring for the prisoners:** Blakey, p. 157.

x **Since that day:** *Encyclopedia of the Civil War*, p. 289.

x **Although it was important:** *Historical Statistics*, V. 1, pp. 24–37.

xi **Even in the spring:** *War of the Rebellion: A Compilation of the Official Records of the Union and Confederate Armies* (henceforth, abbreviated as *O.R.*), Series 2, V. 5, p. 855.

xi **A plume of smoke:** Tobie, pp. 43–46; MacCauley, pp. 66–67.

CHAPTER 1: THE CONFEDERATE CAPITAL

1 **"Death held a carnival":** Sally Putnam, pp. 151, 154.

1 **Indeed, when the war:** DeLeon, pp. 85–86; Hansen, p. 33.

2 **More consequential once:** Thomas, pp. 21–23; Ryan, pp. 8–9; Kimmel, p. 58.

2 **Looming over the scores:** DeLeon, pp. 91–92; Wright, p. 5; Dabney, p. 183; Thomas, p. 23; Howe, p. 550.

3 **Quiet but charged:** Thomas, pp. 33, 44, 73; DeLeon, pp. 85–86; Sally Putnam, pp. xiii–xv, xix–xxi, 76–78, 320.

4 **Laborers, retailers, and contractors:** Sally Putnam, p. 105.

4 **Richmond was transformed:** Thomas, p. 134; Dabney, pp. 160–161; Sally Putnam, p. 208.

4 **No sooner had:** DeLeon, p. 126; Dabney, pp. 163–164, 182; Thomas, pp. 78–81.

5 **Jefferson Davis had respected:** Blakey, pp. i–xv, 6–8, 37, 45, 99.

6   **As provost marshal, Winder:** Dabney, p. 171; Blakey, pp. 51–52, 122–123, 141, 131, 136.

7   **The thudding cannon fire:** Catton, *The Civil War*, p. 64; Bill, pp. 137–138; *Encyclopedia of the Civil War*, pp. 316–317; Sally Putnam, pp. 151, 154.

7   **Manassas had awakened:** Girard, pp. 91–92; Dabney, pp. 175–177; Kimmel, p. 110; Wright, p. 114; Sally Putnam, pp. 210–211; Thomas, p. 58.

8   **Predominantly female and:** John Jones, V. 1, pp. 284–286; Thomas, pp. 119–120; Wright, p. 104; Varon, *Southern Lady*, pp. 102–104.

9   **Having suppressed the riot:** Thomas, pp. 121, 17–19; Blakey, pp. 146–147; Domschke, p. 46; *Daily Richmond Examiner*, April 3–4, 1863; Sally Putnam, pp. 208–209; McGuire, pp. 202–204.

10   **The Confederate government moved:** Wright, p. 104.

10   **City, state, and:** Blair, p. 75; Thomas, pp. 146–147, 169; Blakey, pp. 146–147; John Jones, V. 2, p. 101.

11   **The food shortages:** Thomas, p. 113; Surdam, pp. 98–102; Schwab, pp. 281–282.

11   **To keep a roof:** Sally Putnam, pp. 250–251; McGuire, pp. 196, 245, 247.

12   **Transients of the worst:** Sally Putnam, pp. 76–77; McGuire, p. 203; Kimmel, pp. 50–51, 79, 188.

12   **Surrounded by violence:** *Daily Richmond Examiner*, September 22, 1863; Kimmel, pp. 78, 101, 174, 56.

13   **The old churchgoing Richmond:** Wright, p. 111; Sally Putnam, pp. 253–254; DeLeon, pp. 148–150; Bill, pp. 188–189; McGuire, pp. 328–329; Chesnut, pp. 551, 694; *Richmond Daily Whig*, from Thomas, p. 152.

14   **Saddled, riderless, and:** DeLeon, p. xx; Thomas, p. 124; McGuire, p. 212.

14   **The South's grief was:** McGuire, p. 268; Thomas, pp. 26, 122–123; Kimmel, pp. 154–155; Chesnut, p. 453.

15   **Abraham Lincoln's Emancipation Proclamation:** *Encyclopedia of the Civil War*, p. 112.

15   **Undoubtedly the Confederacy's:** McGuire, p. 268; Kimmel, pp. 154–155; Bill, pp. 200–201.

16   **By autumn 1863:** Sally Putnam, p. 315; John Jones, V. 2, p. 101.

16   **While Richmond's citizens:** Sally Putnam, pp. 102–103; McPherson, *What They Fought For*, pp. 11–12; Thomas, p. 143.

CHAPTER 2: THE POW ARCHIPELAGO

19   **"The arrangement I have made":** O.R., Series 2, V. 5, p. 853.

19   **Francis Lieber's interest:** *Concise Dictionary of American Biography*, p. 567; Lieber, pp. 319–320, 323–333, 337; O.R., Series 3, V. 3, pp. 148–164.

21   **At the beginning:** Reid Mitchell, pp. 572–573.

21   **But by early 1862:** Starkey, pp. 6–7; Vance, pp. 312–315, 6–8; Lloyd, pp. 119–121; O.R., Series 2, V. 6, pp. 266–268; Fischer, p. 376.

22   **At their meeting, Cobb:** Blakey, p. 160; Hesseltine, p. 24; McPherson, *Ordeal by Fire*, p. 451.

22   **And then the spring:** O.R., Series 2, V. 4, pp. 174ff, 210, 239; Hesseltine, pp. 74–77.

23   **The first exchanges occurred:** Hesseltine, pp. 68, 31, 114; Blakey, pp. 156–157; Szabad, p. 80.

23   **In March 1863:** O.R., Series 2, V. 5, p. 853; Hesseltine, p. 69; www.politicalbandwagon.com; *Encyclopedia of Civil War*, p. 332.

24   **Ould the commissioner:** O.R., Series 2, V. 5, p. 855; V. 6, pp. 474–476.

24   **Indeed the Confederacy's:** Casstevens, pp. 8–9; Hesseltine, pp. ix, 96.

24   **The Confederate prisons were:** Casstevens, pp. 5–6; Blakey, p. 213.

25   **Because the Lincoln administration:** McCagg, pp. 478–497; Hesseltine, pp. 35–38, 189, 181; O.R., Series 2, V. 5, p. 489.

26   **A wide gulf lay:** Sturgis, p. 276.

26   **Instead of being systematically:** *Ibid.*, pp. 271–272; Hesseltine, pp. 44, 48–50; Chamberlain, pp. 368–369; O.R., Series 2, V. 4, pp. 152–153; V. 5, pp. 305–306, 317–318.

27   **But Rebel prisoners:** Hesseltine, pp. 183–189, 192; McPherson, *Ordeal*, pp. 450–451; Speer, *Portals*, pp. xiv, 341; Gray, pp. 89, 91–94, 99, 153; Reid Mitchell, p. 565; Blakey, p. 212; *Southern Historical Society Papers*, V. 1, p. 296, letter to *New York World*.

27   **In July 1863:** O.R., Series 4, V. 2, pp. 345–347, for Davis's message to Congress.

28   **While Southern leaders:** O.R., Series 2, V. 5, pp. 940–941; Blakey, p. 160; Sturgis, pp. 297–298.

28 **The resolution struck:** *O.R.*, Series 2, V. 6, p. 528; Lieber Code, Article 58; *The Civil War, A Newspaper Perspective*; McPherson, *Tried By War*, p. 204.

29 **In addition to the:** Moran, *Bastiles*, pp. 4–5; *O.R.*, Series 2, V. 6, pp. 569, 615–618; Reid Mitchell, pp. 574–575; Catton, *The Civil War*, p. 296.

29 **The Confederacy and:** *O.R.*, Series 2, V. 6, pp. 582–583, 594–600; V. 5, p. 469.

30 **Now began a:** *O.R.*, Series 2, V. 6, pp. 532–534.

[Principal Confederate war prisons: Libby, Belle Isle, Castle Pinckney, Castle Thunder, and Ligon's, all in Richmond; Danville and Lynchburg, Virginia; Andersonville, Camp Lawton, Camp Davidson, and Camp Oglethorpe (Macon), Georgia; Cahaba, Alabama; Camp Ford, Texas; Salisbury, North Carolina; Shreveport, Louisiana; and Columbia, Charleston, and Florence, South Carolina.]

[Principal Union war prisons: Johnson's Island, Columbus (state penitentiary), and Camp Chase, Ohio; Camp Morton, Indiana; Alton (state penitentiary), Rock Island, Camp Butler, and Camp Douglas, Illinois; Point Lookout and Fort McHenry, Maryland; Fort Delaware, Delaware; Elmira and Fort Lafayette, New York; Fort Warren, Massachusetts; Old Capitol, Washington, D.C.]

CHAPTER 3: INSIDE LIBBY PRISON

31 **"The whole secret":** Peelle, pp. 70–72.

31 **When Rebel battlefield commanders:** Miller, V. 7, p. 121; Obreiter, p. 178; Parker, p. 22; Morgan, "Libby Warehouse," pp. 36–37; Blakey, pp. 57, 61, 213.

33 **Libby Prison was:** Szabad, pp. 44–45; *Harper's Weekly*, Micro 1–297, V. 7; Gray journal from Ryan, p. 65; Libby Jr., pp. 49–50; Parker, p. 20; Cornwell, pp. 51, 56; Obreiter, p. 178; *O.R.*, V. 6, pp. 544–545, report by Confederate Major Isaac H. Carrington; Miller, V. 7, p. 121.

34 **Colonel Frederick Bartleson described:** Peelle, pp. 70–72.

34 **As in any prison:** Cavada, p. 47; Roach, pp. 72–73; Parker, p. 42; Wilkins, p. 42; Harrold, p. 31; U.S. Congress. House. "Report on the Treatment of Prisoners of War," p. 933; Sneden, V. 5, Part 2, pp. 192, 278; Casstevens, p. 268; Sturgis, p. 287; Abbott, p. 21; George Putnam, pp. 20–21; Szabad, p. 49; Browne, p. 56.

35   **And so the prisoners:** Domschke, pp. 44–45; Bristol, p. 132; Glazier, pp. 67–68; Beaudry, pp. 4–27; Parker, p. 47; Rathborne, p. 45; Heffley, p. 57; Szabad, pp. 44–45, 87; Johnston, pp. 48–49; Cornwell, p. 57; Urquhart; Kent, pp. 7, 38; Roach, p. 75; Jeffrey, pp. 127, 56.

36   **Classes were held:** Bristol, pp. 97, 103–104; Heffley, pp. 60–61; Wells, "The Tunnel Escape," pp. 94–95; Cornwell, pp. 64, 103.

37   **In an era when:** Cavada, p. 53; Beaudry, p. 5; Grant Papers; Cornwell, pp. 114, 126.

37   **Before there were classes:** Peelle, pp. 70–72; Casstevens, p. 268; Sturgis, p. 287; George Putnam, pp. 20–21; Szabad, p. 49; Jeffrey, pp. 10–11, 14; Pierson, *Prisoners Club*; Byrne, "Libby Prison," p. 433; Sanderson, pp. 133–134; Kent, p. 38; Heffley, p. 97.

38   **At ten o'clock on:** Cavada, p. 36; Parker, p. 46; Beaudry, pp. 1–6.

39   **Among the announcements:** Beaudry, pp. 4–27, 45–46; Cornwell, pp. 64, 70, 103, 105.

40   **Vigorous men in their:** Caldwell, pp. 22–24.

41   **Before the Rebels announced:** Beaudry, p. 36; Chamberlain, p. 354; Szabad, p. 88; Bristol, p. 139; Lewis, "Libby," p. 293; Domschke, pp. 57–58.

42   **After lights out:** Roach, p. 79; Domschke, p. 47.

42   **The cry "Letters from home!":** Peelle, p. 42; Johnston, pp. 93–94; Beaudry, p. 12; Keeler, p. 79.

43   **Total candor was impossible:** Parker, p. 45.

43   **The captives skirted the censors:** Isham, p. 179; Dow, pp. 721–722.

43   **Lieutenant George Grant made:** Grant Papers; Fales, p. 16; Glazier, p. 58; Kent, p. 55; Dow, pp. 721–722.

44   **Later in life, some:** Bristol, p. 104; Wyman, p. 27; Caldwell, pp. 28, 19, Dow, pp. 729–730.

44   **Of all the calamities:** U.S. Sanitary Commission, p. 41; Johnston, pp. 56–58; Roach, pp. 58–59; Bliss, pp. 49, 54–57, 62, 64–65; Ryan, p. 92; Rathborne, p. 49.

45   **In the west cellar:** Chamberlain, p. 356; Roach, p. 98.

CHAPTER 4: THE DEFIANT COLONEL

47   **"Even criminals guilty":** *O.R.* Series 2, V. 6, pp. 241–242.

47   **This wasn't how:** Goodnoh, p. 23; Wright, p. 2; Willett, pp. 17–18.

48   **Streight wanted to emulate:** Willett, pp. 18, 11.

49   **Streight's plan was risky:** Willett, pp. 29, 9.

50   **The endurance contest:** Streight report, in *O.R.* Series 1, V. 23, pp. 285–293; Willett, pp. 152–164; Roach, pp. 18–30; Cook, pp. 262–263; Moran, *Bastiles*, pp. 3–4.

50   **As they marched:** Booth, pp. 55–56; *O.R.*, Series 2, V. 6, p. 242; Roach, p. 76.

51   **With the blessing:** Beaudry, p. 16; Moran, *Bastiles*, p. 23; U.S. Sanitary Commission, pp. 155–156; Cornwell, p. 52; Cassteven, p. 265; Szabad, pp. 46–47; Booth, p. 59; Caldwell, p. 5; Urban, p. 297.

51   **Libby Prison's first:** Jeffrey, pp. 86–90; Kimmel, p. 79.

52   **Richard Turner continued:** Oake, pp. 124–125; Di Cesnola, p. 2; Greacen, p. 137.

52   **A Libby captive threw:** Domschke, p. 37; Darby, p. 89; Bristol, pp. 125–126; Thomas P. Turner, Receipts.

53   **It was well worth:** Chamberlain, p. 357; Royall, pp. 54–55ff; Bristol, p. 126; Peelle, p. 70; Byers, pp. 13–14; Sneden, V. 5, Part 2, p. 281.

54   **After the Rebels had:** Roach, pp. 76–77; Shrady, p. 93; Cavada, pp. 43–44, 26.

55   **Confederate officials immediately:** *O.R.*, Series 2, V. 5, pp. 946–947; Roach, p. 45; U.S. Congress. House. "Report on the Treatment of Prisoners of War," pp. 459–460; Browne, p. 269.

55   **The Yankee officers crowded:** Caldwell, p. 12; Bristol, p. 134; Roach, p. 49; Peelle, pp. 85–87.

56   **At one o'clock:** Peelle, pp. 85–87; Caldwell, pp. 12–14; Fales, p. 13; Bristol, pp. 134–135.

57   **And then, two days:** Bristol, pp. 94–96, 192–202.

58   **In April 1863, two:** DeMoss, "Short History."

58   **On July 6, as Libby Prison's:** Lippincott, "Lee-Sawyer Exchange," pp. 39–40; Chamberlain, pp. 358–359; Caldwell, pp. 15–18; *National Intelligencer*, August 15, 1863.

59   **Mrs. Sawyer traveled:** Lippincott, "Lee-Sawyer Exchange," pp. 39–40; *O.R.*, Series 2, V. 6, p. 991; *New York Times*, July 19, 1863.

60   **The adversaries had been:** *O.R.*, Series 2, V. 3, pp. 738–739; Hesseltine, p. 13.

60   **No sooner had Libby Prison's:** Browne, p. 265.

60   **In Richmond, the conflation:** McGuire, pp. 229–230; Sally Putnam, pp. 229, 248, 233; Chesnut, pp. 452–453; *O.R.*, Series 4, V. 2, p. 687.

61 **Days after the twin:** John B. Jones, condensed *Diary*, p. 242.

61 **Amid the military reversals:** Hesseltine, pp. 103–104.

61 **Streight basked in:** Browne, p. 269; Richardson, p. 370; *CSA Henrico*, Micro 590, Frame 912.

62 **The Confederates despised Streight:** *O.R.*, Series 2, V. 6, pp. 241–242.

62 **In the September 11:** Beaudry, p. 20.

63 **In fact, the letter:** Roach, p. 54; Domschke, p. 51; *O.R.*, V. 6, p. 279. [The other signatories were Captain David A. McHolland and Colonel Charles W. Tilden].

63 **A former New York:** Parker, pp. 42–44; Sanderson, pp. xi–xii, xiv, xxii–xxiii; www.zipcon.net/~silas/links.htm; Roach, pp. 55–56; *O.R.*, Series 2, V. 6, pp. 279, 301–303.

CHAPTER 5: MISERY AND RETALIATION

65 **"In the prison cell":** Sturgis, p. 290.

65 **"This may seem":** *O.R.*, Series 1, V. 44, p. 13.

65 **In October 1863:** Di Cesnola, p. 2; Hill, p. 271; Fortescue, p. 114; Tower, p. 3; Booth, p. 72; Murray, p. 33; Sneden, V. 5, Part 2, p. 198; Kent, pp. 33–34; Beaudry, p. 22.

65 **The prisoners demanded:** Domschke, pp. 40–41; Fortescue, p. 108; Browne, pp. 280–282.

66 **It made little difference:** U.S. Sanitary Commission, pp. 156–157, 152; *What Happened When*, p. 421; Isham, pp. 178, 183; Hill, pp. 271–272; Domschke, p. 51; George Putnam, p. 44; Johnston, p. 93; Booth, p. 64; Davidson, p. 54; Dougherty, pp. 9–10, 19–20.

66 **The prisoners scorned:** Johnston, p. 53; Hill, p. 273.

67 **The captives pooled whatever:** Domschke, p. 41; Abbott, p. 25; George Putnam, p. 44.

67 **The last refuge from:** Cornwell, pp. 59–60; Peelle, pp. 70–71.

68 **But when the Confederates:** Cornwell, pp. 57–58; Grant Papers; Prutsman, p. 25; Chamberlain, p. 348; Byers, p. 14; Butler, *Correspondence*, V. 3, p. 435.

68 **Of all the thousands:** *Encyclopedia of American History*, pp. 763, 1016; Byrne, *Prophet of Prohibition*, pp. 3–5, 10, 38, 43, 53–59, 87; Dow, pp. 620, 706–716; Bristol, pp. 138–139; Byers, pp. 13–14; Glazier, p. 59.

69 **In late October 1863:** Dow, pp. 723–724. U.S. Sanitary Commission, pp. 48–51, 158–159; *O.R.*, Series 2, V. 6, pp. 482–483; Sabre, pp. 42–49; Hobart, p. 397; Moran, *Bastiles*, p. 28; Sturgis, p. 305.

70 **After leaving Belle Isle:** *O.R.*, Series 2, V. 6, pp. 482–483, 485.

70 **Ould responded by prohibiting:** *O.R.*, Series 2, V. 6, pp. 522–523; Dow, pp. 724–725.

70 **Unbeknownst to Ould:** *O.R.*, Series 2, V. 6, pp. 510–511.

71 **The potential repercussions:** *Ibid.*, pp. 439, 497.

71 **Fearful that Libby's prisoners:** William Turner, "The Libby Lion," August 1929, p. 30; Booth, p. 76; Moran, pp. 212, 110–111; U.S. Sanitary Commission, p. 152, letter from Captain A. R. Calhoun.

71 **Lieutenant George Forsyth:** William Turner, "The Libby Lion," August 1929, p. 30; U.S. Sanitary Commission, p. 162; Moran, p. 212; Heffley, p. 106; Wilkins, p. 42; Booth, p. 76.

72 **General Meredith requested:** *O.R.*, Series 2, V. 6, pp. 537–538.

72 **On November 26, Meredith:** *O.R.*, Series 2, V. 6, pp. 569–571.

73 **In November 1863:** Heffley, p. 69; *New York Herald*, November 28, 1863; *O.R.*, Series 2, V. 6, pp. 572–574.

73 **Because they were:** Glazier, p. 54; Beaudry, pp. 37–41; *New York Times*, December 5, 1863; Hesseltine, p. 190; *National Intelligencer*, November 11, 1863.

74 **While all of this:** *O.R.*, Series 2, V. 6, pp. 475–476, 535–536.

75 **The first demands for:** Surdam, pp. 62–64; Thomas, pp. 144–145; *O.R.*, Series 4, V. 2, pp. 158–160; Series 2, V. 4, p. 498.

75 **Yet the North remained:** *O.R.*, Series 2, V. 6, pp. 523–524. [Rebelling against British East India Company encroachments, Indians in 1756 crammed 146 British prisoners and their sympathizers into a cell made to hold six to eight men. One hundred twenty-three of them suffocated in the so-called Black Hole of Calcutta.]

76 **Many newspapers advocated:** *New York Times*, November 15, 1863; Blakey, p. 55; *National Intelligencer*, November 23, 1863, reprinting *Boston Daily Advertiser* editorial.

76 **But these arguments:** Sturgis, pp. 280–281; *O.R.*, Series 2, V. 6, pp. 510–514; *Congressional Globe*, 38–2, V. 1, pp. 430–431.

77 **The advocates of retaliation:** Hesseltine, pp. 188–189; *O.R.*, Series 2, V. 6, pp. 486, 625.

77 **At the same time:** *O.R.*, Series 2, V. 6, p. 485; V. 7, pp. 150–151; Speer, p. 126.

77 **Fleeting optimism that an:** *O.R.*, Series 2, V. 6, pp. 313, 315–316; Series 1, V. 44, p. 13; Blakey, p. 18.

78 **As it was understood:** *National Intelligencer*, November 11, 1863.

78 **To counteract the inflammatory:** *O.R.*, Series 2, V. 6, pp. 544–546, 552.

79 **After publishing the report:** *National Intelligencer*, November 24, 1863, reprinting *Richmond Dispatch* editorial of November 13; *Daily Richmond Examiner*, October 29–30, 1863; *Daily Richmond Enquirer*, October 31, 1863.

79 **The whitewash and:** *O.R.*, Series 2, V. 6, p. 843.

80 **No Union general was more:** *O.R.*, Series 2, V. 5, pp. 795–797; *Encyclopedia of the Civil War*, pp. 55–56; Butler, *Autobiography*, pp. 256–259, 374–379, 392–393, 417–418, 425–426, 445. [After the war, Butler assisted the widow and children of the hanged man, William Mumford, by paying off a mechanic's lien on a house built for Mrs. Mumford, and by obtaining a clerk's position for her.]

81 **On December 17, Butler:** *Concise Dictionary of American Biography*, p. 139; *Encyclopedia of the Civil War*, pp. 55–56; Bland, pp. 7–8; Butler, *Correspondence*, V. 3, pp. 148–149; *O.R.*, Series 2, V. 6, pp. 711–712.

81 **The Confederacy strenuously objected:** *O.R.*, Series 2, V. 6, pp. 958–960; Hesseltine, pp. 215–216.

81 **Butler's efforts resulted:** Hesseltine, pp. 210–211, 216, 220–221; *National Intelligencer*, December 21, 1863; *O.R.*, Series 2, V. 6, pp. 769–771.

81 **As hopes for the:** Butler, *Correspondence*, V. 3, p. 435; *O.R.*, Series 2, V. 6, pp. 686, 973–974; Domschke, pp. 52–53; Hesseltine, p. 191; Heffley, p. 94; Szabad, pp. 86, 93–94; Byrne, "Libby Prison," pp. 437–438.

CHAPTER 6: MISS VAN LEW'S SPY RING

83 **"Employ . . . only those":** Butler, *Correspondence*, V. 3, pp. 564–565.

83 **On tiptoe, Annie followed:** Van Lew, Papers, John P. Reynolds Jr. address, p. 3; *New York Times*, October 27, 1900.

84 **As a young woman:** *Spies, Scouts*, p. 86; Varon, "True to the Flag," p. 68; Varon, *Southern Lady*, pp. 9–12; Van Lew, Papers.

84 **In 1836, John Van Lew:** Varon, *Southern Lady*, pp. 9–12, 21–23; Schultz, pp. 48–50; Van Lew, *Yankee Spy*, pp. 4–5, from editor's introduction.

84 **Slavery was so tightly:** Van Lew, Papers; Van Lew, *Yankee Spy*, pp. 4–6, editor's introduction; Varon, *Southern Lady*, p. 14.

85 **As a young adult:** Van Lew, *Yankee Spy*, pp. 5–6; Varon, *Southern Lady*, pp. 24–28; Varon, "True," pp. 68–69.

86 **Although they were pro-Union:** Van Lew, Papers; Varon, *Southern Lady*, pp. 28–29, 35; Ryan, pp. 6, 31.

86 **In August 1861:** Schultz, pp. 50–51; Jeffrey, p. 83; Varon, *Southern Lady*, pp. 64, 68; Ryan, pp. 17–19.

86 **Every day or so:** Varon, *Southern Lady*, pp. 86, 65–66; Schultz, p. 51; Van Lew, *Album*.

87 **Throughout 1861 and in:** Schultz, p. 54.

87 **Van Lew's kindnesses:** Ely, p. 186; Varon, *Southern Lady*, pp. 64–65; *New York Times*, October 22, 1861.

88 **In an attempt to:** Varon, *Southern Lady*, pp. 65, 67; Van Lew, Papers.

88 **By that time:** Van Lew, *Yankee Spy*, pp. 97, 109–110; Markle, p. 182; Varon, "True," pp. 68, 71–72, 74; Varon, *Southern Lady*, pp. 88–92, 94–95, 152, 122, 132; U.S. Congress. House. "Mrs. Abby Green," pp. 1–2.

90 **In March 1862, just as:** Blair, p. 56; Sally Putnam, pp. 101–102; *Concise Dictionary of American Biography*, p. 98; Thomas, p. 82; Blakey, pp. 132–133; Botts letter from *Daily Richmond Examiner* reprinted in *National Intelligencer*, November 26, 1863.

91 **The crackdown had no:** Markle, pp. 143–146; Pinkerton, p. 479; Varon, *Southern Lady*, pp. 74–75.

91 **Winder's crackdown and:** Sally Putnam, pp. 101–102.

92 **The Van Lews' generosity:** *Richmond Dispatch*, July 31, 1861; Van Lew, Papers; Varon, "True," p. 68; Varon, *Southern Lady*, p. 102.

92 **While the Van Lews:** Varon, *Southern Lady*, pp. 77–78; Van Lew, *Yankee Spy*, p. 49; Van Lew, Papers, "Occasional Journal."

93 **One day in 1863:** Domschke, pp. 39–40; Bristol, p. 128; Caldwell, p. 27; Moran, *Bastiles*, p. 213.

93 **On this day, Ross:** Van Lew, *Yankee Spy*, pp. 108–109; Parker, p. 57. [After the war, William Lownsbury sent a box of cigars to Erasmus Ross as a token of his appreciation.]

94 **Erasmus Ross had not:** Van Lew, Papers; Van Lew, *Yankee Spy*, pp. 108–109; Varon, "True," pp. 8, 71, 74.

95 **One day in December 1863:** Varon, *Southern Lady*, pp. 109–111; Butler, *Correspondence*, V. 3, pp. 228–229. [Van Lew's letter did not survive.]

96 **Butler's letter to Elizabeth:** Butler, *Correspondence*, V. 3, p. 319ff; Varon, *Southern Lady*, p. 113.

97 **In early 1864:** Markle, pp. 120, 182.

98 **At first, she merely:** Varon, *Southern Lady*, pp. 163–164, 168, 109–110; Ryan, p. 31; Varon, "True," p. 75; Axelrod, p. 113; Feis, pp. 238–239.

98 **When General Ulysses Grant:** Van Lew, Papers, *Richmond Evening Journal* clip; Varon, *Southern Lady*, pp. 93, 158–159; Schultz, p. 52; Butler, *Correspondence*, V. 3, p. 485; Van Lew, *Yankee Spy*, p. 83.

99 **Butler's connection with Van Lew:** *Spies, Scouts*, pp. 84–86; Butler, *Correspondence*, V. 3, pp. 359, 564–565; Feis, p. 238.

99 **Elizabeth Van Lew went:** Van Lew, Papers; Van Lew, *Yankee Spy*, pp. 3–4.

99 **Increasingly, she was threatened:** Van Lew, Papers, "Occasional Journal"; Van Lew, *Yankee Spy*, pp. 92–93, 97; Varon, *Southern Lady*, p. 254.

100 **In 1864 the Confederate:** Varon, "True," p. 77.

100 **They shockingly underestimated:** Varon, *Southern Lady*, pp. 165–168; Schultz, pp. 52–53; Van Lew, Papers, Reynolds address, p. 5. [During the war, Richards was a Union underground "detective," according to Elizabeth Varon. A 1911 *Harper's Monthly* article reported as fact, without hard evidence, that it was she who infiltrated the Confederacy's inner sanctum.]

101 **In January 1864:** Van Lew, *Yankee Spy*, pp. 59–60; Varon, *Southern Lady*, pp. 130, 132; Schultz, pp. 62–63.

CHAPTER 7: THE WARRIOR SCHOOLTEACHER

103 **"A prisoner, if he deserves":** Cavada, p. 138.

103 **Colonel Thomas Ellwood Rose:** Goodnoh, p. 2; Cavada, p. 160; Moran, "Colonel Rose's Tunnel" in *Famous Adventures*, p. 187; Brown, pp. 10–11; Sneden, V. 5, Part 2, p. 198.

103 **If it all seemed:** Obreiter, p. 170.

104 **Born and raised in:** Interview with Donna Humphrey, Bucks County (Pa.) Historical Society, July 15, 2008; Obreiter, p. 175; Bates, V. 2, pp. 985–989, 992.

104 **After Gettysburg and Vicksburg:** *Encyclopedia of the Civil War*, pp. 73–75; www.civilwarhome.com/Battles.htm.

105 **At nightfall on September 19:** Moran, "Colonel Rose's Tunnel" in *Famous Adventures*, p. 185; Bucks County Historical Society interview; *The Union Army*, V. 1, pp. 413–414; Hanson, pp. 461–462; Arnold, pp. 7, 84; Catton, *The Civil War*, pp. 179–181; Obreiter, p. 230; Peelle, pp. 59–60.

105 **In Richmond, the captives:** Booth, pp. 55–56; Kimmel, p. 75.

106 **Twelve days after:** Moran, "Colonel Rose's Tunnel" in *Famous Adventures*, pp. 190–191.

107 **Rose's wiry-thin new:** Boggs, p. 2; Hamilton, pp. 4–5.

107 **The temporary kitchen's:** Moran, "Colonel Rose's Tunnel" in *Famous Adventures*, pp. 189–190.

108 **At last, they began:** Caldwell, pp. 31–33; Roach, p. 81; Moran, "Colonel Rose's Tunnel" in *Famous Adventures*, pp. 191–193.

109 **One night, Colonel Rose:** Moran, "Colonel Rose's Tunnel" in *Famous Adventures*, pp. 193–195.

110 **A week after Rose's:** *Ibid.*, pp. 195–196.

110 **On the appointed night:** *Ibid.*, pp. 196–199.

111 **Libby buzzed with rumors:** *Ibid.*, pp. 199–200.

111 **The east cellar's inaccessibility:** Cavada, p. 138.

112 **It was probably Hamilton:** Johnston, p. 58; Rose, p. 1; Hamilton, pp. 2–3.

112 **From reveille to lights out:** Moran, "Colonel Rose's Tunnel" in *Famous Adventures*, pp. 201–202; Wells, "The Tunnel Escape," pp. 98–99; Hamilton, p. 3.

113 **On December 19:** Hamilton, p. 3; Rose, p. 2.

113 **When they heard sentries:** Hamilton, p. 3; Rose, p. 2; Moran, "Colonel Rose's Tunnel" in *Famous Adventures*, pp. 202–203.

113 **The excavation became a:** Wells, "The Tunnel Escape," pp. 98–99; Moran, "Colonel Rose's Tunnel" in *Famous Adventures*, p. 202; Hamilton, pp. 4–5.

114 **The completed passageway was:** Virgil Jones, "Libby Prison Break," p. 102; Morgan, p. 31; Moran, "Colonel Rose's Tunnel" in *Famous Adventures*, pp. 203–204; Goodnoh, p. 10.

115 **Clearly, if the fireplace:** Goodnoh, p. 10; *Libby Prison War Museum Catalogue and Program*, p. 8; Hamilton, p. 7.

115 **In the cryptlike:** Rose, p. 2.

115 **Incredibly, despite Rose's:** Morgan, p. 31; Virgil Jones, "Libby Prison Break," p. 97.

116 **The rats were chronic:** Moran, "Colonel Rose's Tunnel" in *Famous Adventures*, p. 205.

116 **Rose and Hamilton were:** Virgil Jones, "Libby Prison Break," pp. 96–97.

117 **With his hat, Hamilton:** Moran, "Colonel Rose's Tunnel" in *Famous Adventures*, p. 204; Morgan, pp. 32–33; Rose, p. 2.

117 **After a few days:** Moran, "Colonel Rose's Tunnel" in *Famous Adventures*, p. 204; Domschke, pp. 50–51.

117 **After carefully screening:** Moran, "Colonel Rose's Tunnel" in *Famous Adventures*, p. 205; Hamilton, p. 7.

118 **Rose formed three:** Rose, p. 2; Moran, "Colonel Rose's Tunnel" in *Famous Adventures*, pp. 205–206.

119 **At first, the digging:** Moran, "Colonel Rose's Tunnel" in *Famous Adventures*, pp. 205–206.

119 **While burrowing through:** *Ibid.*, p. 206; Morgan, pp. 32–33.

CHAPTER 8: TEST OF FAITH

121 **"The feelings of that":** Johnston, p. 63.

121 **The captives began the:** Caldwell, p. 31; Moran, "Colonel Rose's Tunnel" in *Famous Adventures*, p. 204; Wells, "The Tunnel Escape," pp. 99–100.

122 **If it hadn't been:** Krick, pp. 113–118; John Jones, V. 2, pp. 125–127.

122 **They had been at work:** Sclater, p. 9; Moran, "Colonel Rose's Tunnel" in *Famous Adventures*, pp. 206–207.

122 **In the room next:** *O.R.*, V. 6, pp. 507–508; *National Intelligencer*, December 7, 1863.

123 **In November 1863 Streight:** Roach, pp. 88–89; Wells, "The Tunnel Escape," p. 97; *Daily Richmond Examiner*, November 23–24, 1863.

123 **During October, Libby's black:** *Daily Richmond Examiner*, November 23, 1863; Roach, pp. 88–89.

123 **As the appointed hour:** Roach, p. 89; Wells, "The Tunnel Escape," p. 97; Domschke, p. 51; Parker, pp. 42–44.

124 **A few weeks after:** Roach, pp. 89–91; Heffley, p. 95; Earle, p. 269; Sanderson, appendix, p. vii.

124 **Streight acted as he:** Roach, pp. 91–93.

124 **The raider colonel's friends:** *O.R.*, Series 2, V. 6, pp. 279, 301–303; Domschke, pp. 51, 59; *New York Times*, December 13, 1863; Sanderson, p. 19, and appendix, pp. viii, ii–iii, xxx–xxxi.

125 **Rose and Hamilton had:** Moran, "Colonel Rose's Tunnel" in *Famous Adventures*, p. 207; Hamilton, p. 3; Obreiter, pp. 182–183.

125 **Now the conspirators:** Brockett, pp. 280–282; Heffley, pp. 98–99; Varon, *Southern Lady*, pp. 121–122; *Civil War: A Newspaper Perspective*, *New York Herald*, February 15, 1864; Sneden, V. 5, Part 2, p. 299.

126 **No longer able to:** Moran, "Colonel Rose's Tunnel" in *Famous Adventures*, p. 213; Caldwell, p. 27.

127 **The day-and-night excavation:** Moran, "Colonel Rose's Tunnel" in *Famous Adventures*, pp. 207–208.

127 **But the scouts returned:** *Ibid.*; Virgil Jones, "Libby Prison Break," pp. 97–98; Sclater, p. 11.

128 **"All the labor":** Johnston, p. 63.

CHAPTER 9: GENERAL BUTLER'S RAID

129 **"You will see that":** Butler, *Correspondence*, V. 3, pp. 331–332ff.

129 **"We are starving":** Lieutenant Cyrus P. Heffley, diary entry from January 1864.

129 **Hat in hand:** *O.R.*, Series 1, V. 33, pp. 519–520; Varon, *Southern Lady*, pp. 114–115; Butler, *Correspondence*, V. 3, pp. 331–332ff; Van Lew, *Yankee Spy*, pp. 55–56.

131 **Even before General Butler:** Butler, *Correspondence*, V. 3, p. 351.

131 **Van Lew's letter, however:** *Ibid.*, p. 373.

132 **The Northern press, too:** *New York Herald*, October 31, 1863; Schultz, p. 20; *Congressional Globe*, 38–2, pp. 118–119, 316.

132 **General Robert E. Lee had:** *O.R.*, Series 2, V. 6, pp. 438–439.

133 **As Lee was recommending:** Blakey, pp. 170–171.

133 **William Sidney Winder's travels:** *Ibid.*, pp. 175–178.

133 **In Richmond, prices for:** Schwab, pp 281–282, and table, p. 293; Sally Putnam, p. 271; Chesnut, p. 552.

134 **Elizabeth Van Lew set out:** Van Lew, *Yankee Spy*, p. 54; John Jones, V. 2, pp. 154, 156, 135.

134 **The subject of civilian:** John Jones, condensed version, pp. 328–329.

134 **On January 19, someone:** John Jones, V. 2, pp. 132–133; Thomas, p. 155.

134 **The prospect of another:** Sally Putnam, pp. 262–263.

134 **In their rage with:** Thomas, pp. 151–152; Sally Putnam, p. 271; John Jones, V. 2, p. 137.

135 **But these visitors were:** Szabad, p. 49; Roach, pp. 95–97; Thompson, "Escapes from Prison," pp. 150–152; Dow, pp. 716–717; Heffley, p. 97; Byers, pp. 20–21; Sally Putnam, pp. 286–287.

136 **Indeed, the plight of:** Cavada, p. 137; Cornwell, pp. 124–125; Booth, pp. 80–81; Reynolds, p. 394; Peelle, p. 45; Domschke, p. 60.

137 **Cold, hungry, and lice-ridden:** Hill, p. 271; Glazier, p. 64; U.S. Sanitary Commission, p. 156; Domschke, p. 58; Sprague, p. 44; Chamberlain, pp. 364–365; Keeler, p. 69.

137 **Half of January passed:** Hesseltine, p. 124; Cornwell, pp. 63, 65, 68; U.S. Sanitary Commission, p. 152; Sneden, V. 5, Part 2, p. 235; Szabad, p. 90; Glazier, pp. 77–79.

138 **When they lay down:** Heffley, p. 101; Cornwell, pp. 106–107, 97; Booth, p. 80; *O.R.*, Series 2, V. 6, pp. 973, 983–984; Butler, *Correspondence*, V. 3, pp. 435–436; Grant Papers.

138 **Colonel Cavada, a close:** Cavada, pp. 156–157; Goss, p. 28; MacCauley, p. 42; George Putnam, pp. 48–50.

139 **But even for the strong-willed:** Bristol, p. 139.

139 **Those who succumbed to:** Glazier, p. 60; Parker, pp. 49–50; Beaudry, p. 21; Sneden, V. 5, Part 2, p. 237.

139 **Yet, the deprivations:** Byers, p. 28; Cavada, pp. 132–133; Domschke, p. 60.

140 **After nightfall, General Wistar:** Cowles, pp. 21–24; Butler, *Correspondence*, V. 3, pp. 396–397.

140 **In Richmond, the fighting:** Byers, p. 24; Peelle, pp. 31–32; Virgil Jones, p. 100; Cornwell, p. 102.

141 **A few days later, Butler:** *O.R.*, Series 1, V. 33, pp. 144–145; Lincoln, V. 7, p. 59; Butler, *Correspondence*, V. 3, pp. 396–397, 399, 401, 421, 425; Virgil Jones, pp. 21–24, 150; *The Rebellion Record*, pp. 579–580.

CHAPTER 10: THE ORDEAL OF TUNNEL FOUR

143 **"From this time forward":** Moran, "Colonel Rose's Tunnel" in *Famous Adventures*, pp. 221–222.

143 **If they began in:** *Ibid.*, p. 208.

143 **But Rose and Hamilton:** *Ibid.*; Hamilton, p. 4.

144 **Rose, Libby's star digger:** Obrieter, pp. 182–183; Johnston, pp. 65–68, 70–77; Moran, "Colonel Rose's Tunnel" *in Famous Adventures*, pp. 209–210; Rose, p. 2. [Besides Rose and Hamilton, the captives who dug tunnel four were: Major George H. Fitzsimmons, 30th

Indiana; Major B. B. McDonald, 101st Ohio; Captain Terrance Clark, 79th Illinois; Captain John F. Gallagher, 2nd Ohio; Captain W. S. B. Randall, 2nd Ohio; Captain John Lucas, 5th Kentucky; Captain I. N. Johnston, 6th Kentucky; Lieutenant Nineveh S. McKeen, 21st Illinois; Lieutenant David Garbet, 77th Pennsylvania; Lieutenant J. C. Fislar, 7th Indiana Artillery; Lieutenant John D. Simpson, 10th Indiana; Lieutenant John Mitchell, 79th Illinois; and Lieutenant Eli Foster, 30th Indiana. From Moran, "Colonel Rose's Tunnel" in *Famous Adventures*, p. 210.]

145 **There was a sense:** Hopkins, p. 16; Rose, p. 3; *National Tribune*, March 3, 1890; Kent, pp. 45–46.

146 **The new tunnel was now longer:** Hamilton, p. 4; Moran, "Colonel Rose's Tunnel" in *Famous Adventures*, p. 211; Johnston, pp. 68–70; *National Tribune*, March 27, 1890.

146 **The worsening conditions inside:** Morgan, p. 33; Moran, "Colonel Rose's Tunnel" in *Famous Adventures*, pp. 211–212.

147 **Something was different:** Moran, "Colonel Rose's Tunnel" in *Famous Adventures*, pp. 215–216; Caldwell, p. 25; Johnston, pp. 75–77.

147 **After lights-out:** Moran, "Colonel Rose's Tunnel" in *Famous Adventures*, p. 216; Johnston, p. 77.

148 **McDonald's appearance at roll:** Caldwell, pp. 25–26; Moran, "Colonel Rose's Tunnel" in *Famous Adventures*, pp. 216–217.

148 **McDonald had risked severe:** Moran, "Colonel Rose's Tunnel" in *Famous Adventures*, pp. 217–218; Johnston, pp. 77–79, 14–35, 217.

149 **Lacking anything else:** Moran, "Colonel Rose's Tunnel" in *Famous Adventures*, p. 219; Johnston, pp. 81–82.

149 **One day, when Johnston:** Moran, "Colonel Rose's Tunnel" in *Famous Adventures*, pp. 219–220; Johnston, pp. 79–82.

150 **As the tunnelers inched:** Virgil Jones, "Libby Prison Break," pp. 99–100; Johnston, pp. 82–83.

151 **Advancing 5 feet per:** *National Tribune*, March 27, 1890; Johnston, pp. 83–85.

151 **Rose wriggled through:** Moran, "Colonel Rose's Tunnel" in *Famous Adventures*, pp. 218–219; Sclater, pp. 15; Wells, "The Tunnel Escape," p. 98; Johnston, pp. 86–87; *National Tribune*, March 27, 1890; Caldwell, pp. 34–35.

152 **After a few hours:** Moran, "Colonel Rose's Tunnel" in *Famous Adventures*, pp. 221–222; Goodnoh, p. 20; Hopkins, pp. 18–19.

153 **After daytime's gray half-light:** Moran, "Colonel Rose's Tunnel" in *Famous Adventures*, pp. 222–223.

153 **When his cramped limbs:** Hopkins, pp. 19–20; Rose, p. 3; Kent, p. 46; Sclater, p. 16; Moran, "Colonel Rose's Tunnel" in *Famous Adventures*, pp. 223–225; Goodnoh, p. 22.

CHAPTER 11: ESCAPE

157 **"I felt the soft ground":** Wells, "The Tunnel Escape," pp. 103–105.

157 **Never did a day:** Cornwell, p. 103; Wells, "The Tunnel Escape," p. 100.

157 **The captives knew that:** Hobart, p. 398.

158 **Outside Libby's walls:** Krick, pp. 113–118; John Jones, V. 2, p. 145, 147; Long, p. 463.

158 **No prisoner looked forward:** Johnston, pp. 90–91; Roach, pp. 103–105.

159 **Colonel James Sanderson:** Major McDonald account in Sanderson, pp. x–xi, xiv, 48–49.

159 **They might have supported:** Earle, p. 269; Peell, pp. 39–40; Dow, pp. 726–727.

159 **While they waited for:** Caldwell, p. 35; Hamilton, p. 5; Moran, "Colonel Rose's Tunnel" in *Famous Adventures*, p. 225; Chamberlain, p. 362.

160 **At seven o'clock:** Moran, "Colonel Rose's Tunnel" in *Famous Adventures*, p. 225; Morgan, p. 35; Virgil Jones, "Libby Prison Break," pp. 102–103.

161 **On this night:** Roach, pp. 103–105; Caldwell, p. 37.

161 **The Confederates had boarded:** Hobart, pp. 401–402; Earle, pp. 279–283; Lieutenant Leander Williams's account in *National Tribune*, August 30, 1906.

163 **Colonel Rose emerged:** Lewis, p. 297; Sneden, V. 5, Part 2, p. 245.

163 **Rose signaled that the way:** Hamilton, p. 9; Moran, "Colonel Rose's Tunnel" in *Famous Adventures*, p. 225; Earle, p. 282; Johnston, p. 96.

163 **A sentinel stationed:** Lewis, p. 297; Peelle, pp. 39–40.

164 **In the upstairs rooms:** Wells, "The Tunnel Escape," pp. 100–101.

164 **Men were pouring out:** Moran, "Colonel Rose's Tunnel" in *Famous Adventures*, pp. 226–227; Hobart, pp. 401–402.

164 **Amazingly, Lieutenant Frank Moran:** Moran, "Colonel Rose's Tunnel" in *Famous Adventures*, pp. 227–228; Peelle, p. 39; Heffley, p. 100; Roach, p. 101; Caldwell, pp. 35–36.

165 **The prisoners swung around:** Moran, "Colonel Rose's Tunnel" in *Famous Adventures*, pp. 228–231; Caldwell, pp. 35–36; Glazier, p. 85; Wells, "The Tunnel Escape," pp. 103–105; Krick, pp. 113–118.

167 **Back in the sleeping:** Caldwell, p. 36.

167 **In this large exodus:** Goodnoh, p. 23; Nofi, p. 370; Roach, p. 100.

167 **When it was Streight's:** *New York Herald*, March 2, 1864; Szabad, p. 51; Roach, p. 100.

167 **As he waited in:** Virgil Jones, "Libby Prison Break," p. 102.

168 **The breakout had been under way:** Brockett, p. 292.

CHAPTER 12: FLIGHT

169 **"I was overcome":** Van Lew, Papers, "Occasional Journal."

169 **"Our guards now think":** Sneden, V. 5, pp. 315–317.

169 **As they left the:** Hobart, p. 402; Johnston, pp. 97–99; Civil War Richmond, www.mdgorman.com, Wasson letter, March 1, 1884; *National Tribune*, March 27, 1890; Earle, pp. 284–285.

170 **As the eastern sky:** Hamilton, p. 5; *National Tribune*, March 27, 1890; Caldwell, pp. 37–38.

171 **Van Lew had prepared:** Van Lew, *Yankee Spy*, pp. 59–60; Varon, *Southern Lady*, p. 127; Schultze, p. 63.

172 **After completing the morning:** Dow, pp. 727–728; Roach, p. 101; Heffley, p. 100; Peelle, p. 34.

173 **The stunning news that:** Grant Papers; Obreiter, p. 183; Moran, "Colonel Rose's Tunnel" in *Famous Adventures*, p. 232; Wells, "Tunneling Out," p. 324.

173 **Major Thomas Turner did not:** Thomas Turner, Letter to Edith Dabney Tunis Sale; Domschke, p. 64; Roach, p. 106.

173 **Under questioning, the Libby:** Wells, "The Tunnel Escape," p. 107; Glazier, p. 87; Sneden, V. 5, pp. 315–317.

174 **Most of Richmond found out:** February 11–14 editions of *Daily Richmond Examiner*, *Richmond Dispatch*, and the *Daily Richmond Enquirer*, from Library of Virginia microfilm.

175 **While Richmond was reading:** Civil War Richmond, www.md
gorman.com; Moran, "Colonel Rose's Tunnel" in *Famous Adventures*,
pp. 230–231; Moran, *Bastiles*, p. 118.

176 **Lieutenant Melville Small:** *Daily Richmond Enquirer* and *Daily Richmond Examiner* of February 12, 1864; Heffley, p. 101.

176 **The Southern newspaper reports:** *National Intelligencer*, February
16, 1864.

176 **The first escapees entered:** *O.R.*, Series 1, V. 33, pp. 559–560, 565–
566; Butler, *Correspondence*, V. 3, pp. 413–414; Earle, p. 289.

177 **It was an immutable:** Cowles, *Military Atlas of the Civil War*, pp. 80–
81; *O.R.*, Series I, V. 33, p. 566; Urquhart.

178 **Dogs were as much:** Caldwell, pp. 37–49; Johnston, pp. 97–104;
Hobart, pp. 402–406; Wallber, p. 127.

178 **Caldwell's luck continued:** John Jones, V. 2, pp. 125, 127; Caldwell,
pp. 37–49.

179 **During their fourth night:** Wells, "The Tunnel Escape," pp. 108–
109; McCreery, pp. 22–28; Caldwell, pp. 49–59; Johnston, pp. 104–
120; Wallber, p. 130; Moore, p. 375.

180 **Slaves and free blacks:** Butler, *Correspondence*, V. 3, p. 468; *Harper's
Weekly*, V. 8, p. 173.

181 **Lieutenant James Wells:** Wells, "The Tunnel Escape," pp. 105–109;
National Tribune, March 27, 1890.

182 **Of all the officers:** Roach, pp. 109–112; Domschke, p. 66.

183 **Colonel Streight, in fact:** U.S. Congress. House, "Mrs. Abby
Green," pp. 1–2; Varon, *Southern Lady*, pp. 128–129.

183 **Six days after the breakout:** Van Lew, Papers, "Occasional Journal";
Van Lew, *Yankee Spy*, p. 63.

184 **Within hours, Confederate detectives:** Varon, *Southern Lady*, pp.
130–131.

184 **Troubled by the:** Van Lew, *Yankee Spy*, p. 63.

184 **On the day that:** *O.R.*, Series 1, V. 33, p. 565.

185 **Streight's friends believed:** *Daily Richmond Examiner*, February 16,
1864; *Richmond Sentinel*, February 20, 1865, from Civil War Richmond, www.mdgorman.com.

185 **Eight days after he wriggled:** Krick, pp. 113–118; Van Lew, Papers,
"Occasional Journal."

185 **Carrying weapons and:** Roach, pp. 113–117; *National Intelligencer*,
March 1, 1864.

187 **Of the 109 officers:** Moran, "Colonel Rose's Tunnel" in *Famous Adventures*, p. 234; *Richmond Dispatch*, February 15, 1864; Lewis, pp. 299–300.

187 **Still, the recaptured escapees:** Ryan, pp. 111–112; Glazier, p. 87; Heffley, p. 101.

188 **Lieutenant Latouche's discovery:** U.S. Congress. House, "Report on the Treatment of Prisoners of War," p. 233.

189 **As they awaited the denouement:** Peelle, p. 37.

189 **Rose was traveling alone:** Hamilton, p. 5.

189 **The Confederate guard released:** Moran, "Colonel Rose's Tunnel" in *Famous Adventures*, pp. 234–236.

189 **The next night, while wading:** *Ibid.*, pp. 237–242; Goodnoh, p. 30.

CHAPTER 13: FALLOUT

193 **"We will cross":** *O.R.*, Series 1, V. 33, pp. 178–179.

193 **"Major Thomas Turner shook up":** Turner, Special Instructions; Heffley, pp. 101–102; Szabad, pp. 101–102; Cavada, pp. 174, 189.

193 **Workmen installed a:** Peelle, p. 52; Parker, p. 63; Szabad, p. 52; Domschke, p. 67; Byers, p. 26; George Putnam, pp. 27–29; Turner, Special Instructions; Keeler, p. 102; Speer, *Portals*, p. 207.

194 **Spencer Deaton of the:** Cornwell, p. 107; Van Lew, *Yankee Spy*, p. 65.

195 **In a letter to President Lincoln:** *O.R.*, Series 2, V. 6, pp. 966, 977–978.

195 **After listening to Colonel:** Butler, *Correspondence*, V. 3, p. 417; *O.R.*, Series 2, V. 6, pp. 979–980.

195 **Lieutenant Morton Tower:** Tower, p. 10.

196 **"Bearing the bandages, water and sponge,":** Whitman, pp. xxvii, 221.

196 **No escapee made a greater:** *New York Herald*, March 2, 1864; Szabad, p. 52.

196 **Streight sat down:** Moore, *The Rebellion Record*, Streight report, March 2, 1864, pp. 450–452.

197 **In response to Streight's:** *New York Times*, March 31, 1864.

197 **While he had the attention:** Hesseltine, pp. 194–195.

198 **And so, when Sanderson:** Sanderson, pp. 8–16, 136–138.

198 **On February 11:** Lincoln, V. 7, p. 178; Schultz, pp. 67–77.

199 **Kilpatrick wanted to lead:** *O.R.*, Series 1, V. 33, pp. 170–171, 162–163, 70–71; Virgil Jones, *Eight Hours Before Richmond*, pp. 25–29; Lincoln, V. 7, pp. 53–56; Schultz, pp. 78, 103–104.

200 **A lot was expected:** Schultz, p. 91; Virgil Jones, *Eight Hours Before Richmond*, p. 152.

200 **On the night of February 28:** Schultz, pp. 91, 100; *New York Times*, March 9, 1864.

201 **There, the raiders rendezvoused:** Schultz, pp. 110–118; George Putnam, p. 277; *Daily Richmond Examiner*, March 5, 1864; John Jones, V. 2, p. 162; *Confederate Veteran*, No. 31, pp. 177–178; *Southern Historical Society Papers*, V. 34, pp. 353–358.

202 **The distant sound:** Schultz, pp. 124–125.

202 **Kilpatrick's resolve began to:** *Ibid.*, pp. 124–125, 128, 132–133.

203 **The alarm bell rang:** McGuire, p. 255; Schultz, p. 128; Sally Putnam, p. 283.

203 **But General John Winder:** Ryan, p. 117; Glazier, p. 89.

203 **Winder, desperate to prevent:** John Jones, V. 2, p. 164; Schultz, pp. 130–131; *New York Times*, March 13, 1864.

204 **With picks and shovels:** Glazier, p. 93; Cornwell, p. 116; Heffley, p. 103; U.S. Sanitary Commission, pp. 42–43, 160.

204 **Kilpatrick's loss of nerve:** *O.R.*, Series 2, V. 8, p. 344.

205 **Dahlgren's five hundred troopers lost:** *O.R.*, Series 1, V. 33, pp. 170–171, 205–206; Schultz, pp. 134–139; Varon, *Southern Lady*, p. 138.

205 **While ransacking Dahlgren's body:** *Richmond Sentinel*, *Daily Richmond Examiner*, March 5, 1864; Davis, p. 543; *O.R.*, Series 1, V. 33, pp. 178–179, 219–220; Schultz, pp. 155–156; Van Lew, *Yankee Spy*, pp. 74–75.

206 **The Union command denied:** Varon, *Yankee Spy*, p. 141; Schultz, pp. 249–250; *O.R.*, Series 1, V. 33, pp. 175–176; Virgil Jones, *Eight Hours Before Richmond*, pp. 137–138, 174.

206 **Genuine or counterfeit:** Schultz, pp. 181–182; Varina Davis quote from her *Memoirs*, V. 2, p. 472, found in Thomas, p. 159.

207 **After rifling through Dahlgren's possessions:** U.S. Congress. "Report on Returned Prisoners," pp. 16–17; Schultz, p. 176.

207 **In Richmond, the coffin:** Schultz, p. 173; *Daily Richmond Examiner*, March 7, 1864; Varon, *Southern Lady*, p. 142.

207 **Now it was the Union's:** Butler, *Correspondence*, V. 3, p. 509; Varon, *Southern Lady*, p. 157; *Daily Richmond Examiner*, March 14, 1864.

208 **Elizabeth Van Lew had been:** Varon, *Southern Lady*, pp. 143–148; Van Lew, Papers, "Personal Narrative"; Van Lew, *Yankee Spy*, pp. 68–72, 82; Varon, "True," p. 74; Schultz, p. 259. [After the war, Union detectives recovered Dahlgren's ring, pocket watch, coat, and prosthetic leg. In June 1865, Dahlgren's body was exhumed and shipped to Washington, where his father identified the remains. Henry Ward Beecher delivered the eulogy at an impressive memorial service attended by President Andrew Johnson and military officials. Dahlgren's final resting place is North Hill Cemetery in Philadelphia.]

208 **Judged a failure:** *O.R.*, Series 1, V. 33, pp. 217–218, 222–223.

209 **Eight officers and a:** Abbott, pp. 257–259; Isham, pp. 33–34; Kent, pp. 35–36; Glazier, pp. 92, 94.

209 **Although the *Richmond Dispatch*:** *Richmond Dispatch*, March 4, 1864; Schultz, p. 150; *O.R.*, Series 1, V. 33, pp. 199–200; Virgil Jones, *Eight Hours Before Richmond*, p. 165; Chesnut, p. 578; Sally Putnam, p. 283.

210 **Hastily convened in response:** *O.R.*, Series 2, V. 7, pp. 110–111; U.S. Congress. "Report on Returned Prisoners," pp. 1–8, 20, 24; Hesseltine, pp. 196–197; *Congressional Globe*, 38–2, V. 1, p. 268; U.S. Congress. House. "Report on the Treatment of Prisoners of War."

210 **Three months later:** U.S. Sanitary Commission, pp. 22–26, 72–74, 92–99, 197.

211 **In the early spring:** Dow, pp. 731–733; Heffley, p. 105; Sally Putnam, p. 285.

211 **Prisoner exchanges had resumed:** Grant, *Memoirs*, pp. 364–366; Speer, *Portals*, p. 15; Butler, *Autobiography*, pp. 592–595; *O.R.*, Series 2, V. 7, pp. 50, 62–63, 615; U.S. Congress. Senate. "Report of the Joint Committee on the Conduct of the War," pp. 76–77.

212 **Yet except for the:** *New York Times*, December 27, 1864.

213 **When Butler entered politics:** Butler, *Autobiography*, p. 610.

213 **As Grant's spring offensive:** Heffley, pp. 108–109; Byrne, "Libby Prison," p. 441.

213 **Belle Isle's 10,000 cadaverous:** Schultz, p. 52; Roach, pp. 62–63; Hesseltine, p. 124; Varon, *Southern Lady*, p. 133; Peelle, p. 82.

214 **Although it had opened:** Sturgis, pp. 312, 291; Boggs, pp. 19–20, 33, 37, 43; *O.R.*, Series 2, V. 7, pp. 546–552; Moran, *Bastiles*, pp. 64–65; Blakey, pp. 182–186.

214 **In June, General John Winder:** Blakey, pp. xii, 1, 158–159, 178–179, 182–184, 188, 197, 201; Heffley, p. 89; Sturgis, p. 291; Boggs, pp. 37–40.

215 **With each passing month:** *Congressional Globe*, 38–2, V. 1, pp. 428–429, 455; *Southern Historical Society Papers*, V. 1, pp. 121, 126–129; V. 4, p. 117; *O.R.*, Series 2, V. 7, pp. 607, 687–691.

216 **As Libby's population waxed:** *O.R.*, Series 2, V. 8, pp. 337–353, Confederate Congress report; Grant Papers.

216 **In January 1865, the dying:** Butler, *Autobiography*, pp. 610–612.

216 **Figures released immediately after:** *Southern Historical Society Papers*, V. 1, pp. 123–124; Boggs, p. 63; Catton, "Prison Camps."

217 **In 1876, eleven years:** *Confederate View*, pp. 115–116; *Southern Historical Society Papers*, V. 1, pp. 116–118.

EPILOGUE

219 **"When a people all think":** Van Lew, Papers.

219 **On the day that Richmond:** *Walls That Talk*, p. 4; Sandra Parker, p. 66.

219 **As the Rebels were abandoning:** Dabney, pp. 189–193; Van Lew, Papers; Van Lew, *Yankee Spy*, p. 105.

220 **Confederate soldiers visited:** David Parker, p. 55; Van Lew, *Yankee Spy*, p. 110; Varon, *Southern Lady*, p. 193.

220 **Unaware that the Rebels:** David Parker, pp. 55–56.

221 **Two days later, President:** Ryan, p. 131.

221 **When Ulysses Grant:** Varon, *Southern Lady*, p. 196; *Spies, Scouts*, p. 89.

221 **After reducing the documents:** Thomas Turner, letter to S. R. Shinn, on January 8, 1866, in *New York Times*, July 7, 1895; Ryan, p. 139.

222 **Libby's hated jailkeeper:** Brown, pp. 25–26; Sandra Parker, pp. 67–68; William Turner, pp. 23–34; Blakey, pp. 202–203; *Richmond Dispatch*, December 7, 1901.

223 **Two months after Rebel:** Moran, "Colonel Rose's Tunnel" in *Famous Adventures*, p. 242; Keeler, p. 104; Cornwell, p. 137; Heffley, p. 108; Bates, V. 2, pp. 990–992; *The Union Army*, V. 1, p. 414; Obreiter, p. 230; Boatner, p. 708; interview with Donna Humphrey; Rose gravesite, Arlington Cemetery.

223 **Neither General Neal Dow:** Willett, p. 18; Byrne, *Prophet of Prohibition*, pp. 105–106.

224 **Libby Prison reverted:** Klee, pp. 33–38; Sandra Parker, pp. 69–71; Urquhart; *Libby Prison War Museum Catalogue and Program*, Duke University Special Collections; Ryan, pp. 133–134; *Walls That Talk*; George Putnam, p. 26.

225 **For more than a century:** Ryan, p. 133.

226 **The war had cost:** Varon, *Southern Lady*, p. 86; Van Lew, Papers, January 1867; Van Lew, Album, pp. 2–2A; Van Lew, *Yankee Spy*, pp. 114–116.

226 **The War Department did:** Van Lew, *Yankee Spy*, p. 110.

226 **Ulysses Grant never forgot:** Varon, *Southern Lady*, pp. 216–218, 228; Van Lew, Papers, October 27, 1876; Varon, "True," p. 80.

227 **Several years later:** Varon, *Southern Lady*, pp. 240–241.

227 **There, she discovered that:** Varon, *Southern Lady*, p. 246; Van Lew, *Yankee Spy*, pp. 129, 131, 111.

228 **During the war, the observant:** Van Lew, Papers; Varon, *Southern Lady*, p. 113; Beymer, pp. 86–99.

228 **A few months later:** *New York Times*, October 27, 1900.

228 **The Virginia Club moved:** *National Tribune*, July 4, 1901, from Civil War Richmond Web site, www.mdgorman.com; Van Lew, Papers; Varon, *Southern Lady*, p. 257.

229 **In 1902, Elizabeth Van Lew's:** Varon, *Southern Lady*, pp. 251–253; Wright, p. 206; Beymer, p. 86; Shockoe Cemetery gravesite.

BIBLIOGRAPHY

Abbott, John S. C. "The Capture, Imprisonment, and Escape." *Harper's Monthly Magazine* 34 (January 1867): 150–170.

Armstrong, William M. "Libby Prison: The Civil War Diary of Arthur G. Sedgwick." *Virginia Magazine of History and Biography* 71 (October 1963): 449–460.

Arnold, James R. *Chickamauga 1863: The River of Death*. Oxford: Osprey, 1992.

Aubery, Cullen B. "Doc." *Recollections of a Newsboy in the Army of the Potomac, 1861–1865*. Milwaukee, Wis., 1904.

Axelrod, Alan. *The War Between the Spies*. New York: Atlantic Monthly Press, 1992.

Bates, Samuel P. *History of the Pennsylvania Volunteers, 1861–1865*. 5 vols. Harrisburg, Penn.: B. Singerly, 1869.

Beaudry, Louis N., ed. *The Libby Chronicle: Devoted to Facts and Fun*. Albany, N.Y.: C. F. Williams Printing Company, 1889. First published in 1863.

Beymer, William Gilmore. "Miss Van Lew." *Harper's Monthly Magazine* 123 (August 1911): 86–99.

Bill, Alfred Hoyt. *The Beleaguered City: Richmond, 1861–1865*. New York: Alfred A. Knopf, 1946.

Blair, Williams. *Virginia's Private War: Feeding Body and Soul in the Confederacy, 1861–1865*. Oxford and New York: Oxford University Press, 1998.

Blakey, Arch Frederic. *General John H. Winder, C.S.A.* Gainesville: University of Florida Press, 1990.

Bland, T. A. *Life of Benjamin F. Butler*. Boston: Lee and Shepard, 1879.

Bliss, George N. *Cavalry Service with General Sheridan, and Life in Libby Prison*. Providence: Historical Society of Rhode Island, 1884.

Boatner, Mark Mayo III. *The Civil War Dictionary*. New York: David McKay Company, 1959.

Boggs, S. S. *Eighteen Months a Prisoner Under the Rebel Flag*. Lovington, Ill.: self-published, 1889.

Booth, Benjamin F., and Steve Meyer. *Dark Days of the Rebellion: Life in Southern Military Prisons*. Garrison, Iowa: Meyer, 1995.

Botzenhart, Manfred. "French Prisoners of War in Germany, 1870–71." In Förster and Nagler's *On the Road to Total War*, 587–693.

Bray, John. "My Escape from Richmond." *Harper's Monthly Magazine* 28 (April 1864): 662–665.

Bristol, Frank Milton. *The Life of Chaplain McCabe, Bishop of the Methodist Episcopal Church*. New York: Eaton & Mains, 1908.

Brockett, L. P. *The Camp, the Battle Field, and the Hospital, or Lights and Shadows of the Great Rebellion*. Philadelphia and Chicago: National Publishing Company, 1866.

Brown, Daniel Patrick. *The Tragedy of Libby and Andersonville Prison Camps*. Ventura, Calif.: Golden West Historical Publications, 1980.

Browne, Junius Henri. *Four Years in Secessia*. New York: Arno & The New York Times, 1970. First published in 1865 by O. D. Case.

Burrows, J. L. "Recollections of Libby Prison." In *Southern Historical Society Papers*, vol. 11, 83–92.

Burson, William. *A Race for Liberty: My Capture, Imprisonment, and Escape*. Wellsville, Ohio: W. C. Foster, 1867.

Butler, Benjamin F. *Autobiography and Personal Reminiscences of Major-General Benjamin F. Butler*. Boston: A. M. Thayer & Co., 1892.

——————. *Private and Official Correspondence of General Benjamin F. Butler During the Period of the Civil War*. 5 vols. Privately issued, 1917.

Byers, S. H. M. *What I Saw in Dixie, or Sixteen Months in Rebel Prisons*. Dansville, N.Y.: Robbins & Poore, 1868.

Byrne, Frank. "A General Behind Bars." *Civil War History* 8 (June 1962): 164–183.

——————. "Libby Prison, a Study in Emotions." *Journal of Southern History* 24 (1958): 430–444.

——————. *Prophet of Prohibition: Neal Dow and His Crusade*. Madison: State Historical Society of Wisconsin, 1961.

Caldwell, David S. *Incidents of War and Southern Prison Life*. Wyandot: Ohio Genealogical Society, 2002. First published in 1864.

Carruth, Gordon. *What Happened When: A Chronology of Life & Events in America*. New York: Signet, 1989.

Casstevens, Frances H. *George W. Alexander and Castle Thunder: A Confederate Prison and Its Commandant*. Jefferson, N.C., and London: McFarland & Company, 2004.

————. *Out of the Mouth of Hell: Civil War Prisons and Escapes.* Jefferson, N.C., and London: McFarland & Company, 2005.

Catton, Bruce. *The Civil War.* Boston: Houghton Mifflin, 1978. First published in 1960 as *The American Heritage Picture History of the Civil War* in New York by American Heritage.

————. "Prison Camps of the Civil War." *American Heritage Magazine* 10 (August 1959): 17–30.

Cavada, Federico F. *Libby Life: Experiences of a Prisoner of War in Richmond, Virginia, 1863–1864.* Philadelphia: King & Baird, 1864.

The Century Illustrated Monthly Magazine. 23–58 (1881–1889). New York: Century Company.

Chamberlain, J. W. "Scenes in Libby Prison." In *Military Order of the Loyal Legion of the United States,* vol. 2, 342–370.

Chesnut, Mary Boykin. *Mary Chesnut's Civil War.* Edited by C. Vann Woodward. New Haven: Yale University Press, 1981. First published in 1905 as *Diary from Dixie* in New York by D. Appleton.

The Civil War: A Newspaper Perspective. Davis Library Reference Department, University of North Carolina. CD-ROM, nos. 10–58. At www.accessible.com.

The Civil War Society's Encyclopedia of the Civil War. New York: Wings Books, 1997.

Civil War Times Illustrated. April 1962–August 2002. Gettysburg, Penn.: Historical Times.

Concise Dictionary of American Biography. 2nd ed. New York: Charles Scribner's Sons, 1977.

Confederate States of America. Department of Henrico Papers, 1861–1864. Virginia Historical Society, Richmond.

Confederate Veteran 1–40 (1893–1932). Davis Library, University of North Carolina. Microfilm, serials 1–423.

Congressional Globe. Washington, D.C.: Blair & Rives, 1834–1873.

Cook, James F. "The 1863 Raid of Abel D. Streight: Why It Failed." *Alabama Review* 22, no. 4 (October 1969): 254–269.

Cooper, Alonzo. *In and out of Rebel Prisons.* Oswego, N.Y.: R. J. Oliphant, 1888.

Cornwell, Robert Thompson. *Libby Prison and Beyond: A Union Staff Officer in the East, 1862–1865.* Edited by Thomas M. Boaz. Shippenburg, Penn.: Burd Street Press, 1999.

Cowles, Calvin D., ed. *The Official Military Atlas of the Civil War.* Washington, D.C.: Government Printing Office, 1891–1895.

Dabney, R. H. "Prisoners of the Civil War." In *Southern Historical Society Papers*, vol. 17, 378–381.

Dabney, Virginius. *Richmond: The Story of a City*. Garden City, N.Y.: Doubleday & Company, 1976.

Daily National Intelligencer. April 1861–April 1865. Washington, D.C.: Gales & Seaton. Library of Congress photoduplication service copy available at North Carolina State University, D. H. Hill Library. Microfilm AN2–D362.

Daily Richmond Enquirer. 1861–1865. Library of Virginia, Microfilms 23, 23b, and 23d.

Daily Richmond Examiner. 1861–1865. Library of Virginia, Microfilms 366, 366a, and 2330.

Darby, George. *The Civil War Memoirs of Sergeant George W. Darby, 1861–1865*. Edited by Rogan H. Moore. Bowie, Md.: Heritage Books, 1999.

Davidson, H. M. *Fourteen Months in Rebel Prisons*. Milwaukee: Daily Wisconsin Printing House, 1865.

Davis, William C. *Jefferson Davis: The Man and His Hour*. New York: HarperCollins, 1991.

DeLeon, T. C. *Four Years in Rebel Capitals: An Inside View of Life in the Southern Confederacy*. Mobile, Ala.: Gossip Printing Company, 1892.

DeMoss, John C. "A Short History of the Soldier Life, Capture, and Death of William Francis Corbin, Captain Fourth Kentucky Cavalry, C.S.A." 1897. At www.rootsweb.ancestry.com/~kycampbe/military.htm.

Di Cesnola, Louis Palma. *Ten Months in Libby Prison*. Philadelphia, 1865. Duke University Special Collections Library.

Domschke, Bernhard. *Twenty Months in Captivity: Memoirs of a Union Officer in Confederate Prisons*. Rutherford, Madison, and Teaneck, N.J.: Fairleigh Dickinson University Press, 1987.

Dougherty, Michael. *Prison Diary of Michael Dougherty*. Bristol, Penn.: Chas. A. Dougherty, 1908.

Dow, Neal. *The Reminiscences of Neal Dow. Recollections of Eighty Years*. Portland, Maine: Evening Express Publishing Company, 1898.

Dozier, Graham T., ed. *Virginia's Civil War: A Guide to Manuscripts at the Virginia Historical Society*. Richmond: Virginia Historical Society, 1998.

Drake, J. Madison. *Fast and Loose in Dixie*. New York: Author's Publishing Company, 1880.

Drummond, Robert Loudon. *The Religious Pray, the Profane Swear. A Civil War Memoir. Personal Reminiscences of Prison Life During the War of the Rebellion*. Edited by Victor E. Taylor. Aurora, Colo.: Davies Group, 2002.

Dumas, Alexander. *The Count of Monte Cristo.* New York and London: Penguin Books, 1996.

Dyer, Brainerd. "Treatment of Colored Troops by the Confederates." *Journal of Negro History* 20 (July 1935): 273–286.

Earle, Charles Warrington. "In and out of Libby Prison." In *Military Order of the Loyal Legion of the United States,* vol. 10, 247–292.

Ely, Alfred. *Journal of Alfred Ely, a Prisoner of War in Richmond, 1862.* New York: D. Appleton and Company, 1862. University of North Carolina Law School, Microform, no. 465.

Estabrooks, Henry L. *Adrift in Dixie: A Yankee Officer Among the Rebels.* New York: Carleton, 1866.

Fales, James M. *Prison Life of Lieut. James M. Fales.* Providence, R.I.: N. Bangs Williams & Company, 1882.

Famous Adventures and Prison Escapes of the Civil War. New York: Century Company, 1898.

Feis, William B. *Grant's Secret Service: The Intelligence War from Belmont to Appomatox.* Lincoln and London: University of Nebraska Press, 2002.

Fischer, David Hackett. *Washington's Crossing.* Oxford and New York: Oxford University Press, 2004.

Foote, Shelby. *The Civil War: Red River to Appomatox.* New York: Vintage Books, 1986. First published in 1974.

Förster, Stig, and Jörg Nagler, eds. *On the Road to Total War: The American Civil War and the German Wars of Unification, 1861–1871.* Cambridge: Cambridge University Press, 1997.

Forsythe, John W. *Guerrilla Warfare and Life in Libby Prison.* Annandale, Va.: Turnpike Press, 1967. First published in 1892.

Fortescue, Louis R. Diary, 1863–4. Southern History Collection, University of North Carolina at Chapel Hill.

Friedel, Frank. *Francis Lieber, Nineteenth-Century Liberal.* Baton Rouge: Louisiana State University Press, 1947.

Girard, Charles. *A Visit to the Confederate States of America in 1863.* Tuscaloosa, Ala.: Confederate Publishing Company, 1962.

Glazier, Willard W. *The Capture, the Prison Pen, and the Escape, Giving a Complete History of Prison Life in the South.* Hartford, Conn.: H. E. Goodwin, 1868.

Goodnoh, E. C. *The Famous Tunnel Escape from Libby Prison: "From Rat Hell to Liberty."* Chicago, 1900. Library of Virginia. Microfiche 157, unit 3, IL27.

Gorman, Michael D. Excellent photographs of Libby Prison and newspaper accounts, official reports, magazine articles, and book excerpts about life inside. At www.mdgorman.com/Prisons/Libby/libby_prison.htm.

Goss, Warren Lee. *The Soldier's Story of His Captivity at Andersonville, Belle Isle, and Other Rebel Prisons.* Boston: Lee and Shepard, 1867.

Grant, George W. George W. Grant Papers, 1861–1892. Duke University Special Collections Library.

Grant, Ulysses S. *Personal Memoirs of U. S. Grant.* Edited by E. B. Long. New York: Da Capo Press, 1982. Reprint of 1952 edition.

Greacen, James. "Fifteen Months in Rebel Prisons." In *The National Tribune Scrap Book*, vol. 3, 134–144.

Grey, Michael P. *The Business of Captivity. Elmira and Its Civil War Prison.* Kent, Ohio: Kent State University Press, 2001.

Hamilton, A. G. *Story of the Famous Tunnel Escape from Libby Prison.* S. S. Boggs, 1895. Duke University Special Collections Library.

Harlow, Reuben. Letter to wife from Libby Prison. December 30, 1863. Virginia Historical Society.

Harper's Monthly Magazine 1–319, nos. 1–1912 (1850–2009).

Harper's Weekly 7 and 8 (1863–1864). Davis Library, University of North Carolina at Chapel Hill. Microfilm, serial 1–297, reel 3.

Harrold, John. *The Capture, Imprisonment, Escape and Rescue of John Harrold, a Union Soldier in the War of Rebellion.* Atlantic City, N.J.: Daily Union Book and Job Printing Office, 1892.

Heffley, Albert, and Cyrus P. Heffley. *Civil War Diaries of Capt. Albert Heffley and Lt. Cyrus P. Heffley, Company F—142nd Regt. Penna. Vol., Army of the Potomac.* Apollo, Penn.: Closson Press, 2000.

Hesseltine, William Best. *Civil War Prisons: A Study in War Psychology.* Columbus: Ohio State University Press, 1930.

Hill, John F. "To and from Libby Prison." In *The Civil War in Song and Story*, edited by Frank Moore, 270–281. New York: P. F. Collier, 1889.

Historical Statistics of the United States, Colonial Times to 1970. 2 vols. Washington, D.C.: U.S. Bureau of the Census, 1975.

Hobart, Harrison C. "Libby Prison—The Escape." In *Military Order of the Loyal Legion of the United States*, vol. 46., 394–409.

Hopkins, Tighe. *The Way out of Libby.* Fort Wayne, Ind.: Public Library of Fort Wayne and Allen County, 1955.

Humphrey, Donna. Bucks County Historical Society. Interview on July 15, 2008.

Hyde, Solon. *A Captive of War: Solon Hyde, Hospital Steward, 17th Regiment Ohio Volunteer Infantry*. Shippensburg, Penn.: Burd Street Press, 1996.

Isham, Asa B., Henry M. Davidson, and Henry B. Furness. *Prisoners of War and Military Prisons: Personal Narratives of Experience in the Prisons at Richmond, Danville, Macon, Andersonville, Savannah, Millen, Charleston, and Columbia*. Cincinnati, Ohio: Lyman & Cushing, 1890.

Jackson, Robert. *The Prisoners, 1914–1918*. London, New York: Routledge, 1989.

Jeffrey, William H. *Richmond Prisons, 1861–1862*. St. Johnsbury, Vt.: Republican Press, 1893.

Johnston, I. N. *Four Months in Libby, and the Campaign Against Atlanta*. Cincinnati, Ohio: Methodist Book Concern, 1864.

Jones, J. William, ed. *Confederate View of the Treatment of Prisoners*. Richmond, Va.: Southern Historical Society, 1876.

Jones, John B. *A Rebel War Clerk's Diary*. Condensed version. Edited by Earl Schenck Miers. New York: Sagamore Press, 1958.

———. *A Rebel War Clerk's Diary at the Confederate States Capital*. 2 vols. New York: Old Hickory Bookshop, 1935.

Jones, Virgil Carrington. *Eight Hours before Richmond*. New York: Henry Holt and Company, 1957.

———. "Libby Prison Break." *Civil War History*. Vol. 4 (June 1958): 93–104.

Journal of Negro History. 1–86 (1916–2001). Washington, D.C.: The Association for the Study of Negro Life and History.

Journal of Southern History. 1–74 (1935–2008). Baton Rouge, La.: Southern Historical Association.

Keeler, Alonzo M. *A Guest of the Confederacy: The Civil War Letters and Diaries of Alonzo M. Keeler, Captain, Company B, Twenty-Second Michigan Infantry*. Edited by Robert D. Allen and Cheryl J. Allen. Nashville, Tenn.: Cold Tree Press, 2008.

Kent, Will Parmiter. *The Story of Libby Prison. Also Some Perils and Sufferings of Certain of Its Inmates*. Chicago: Libby Prison War Museum Association, 1890.

Kimmel, Stanley. *Mr. Davis's Richmond*. New York: Coward-McCann, Inc., 1958.

Kiner, F. F. *One Year's Soldiering*. Prior Lake, Minn.: Morgan Avenue Press, 1863.

Klee, Bruce. "They Paid to Enter Libby Prison." *Civil War Times Illustrated* 37 (February 1999): 32–38.

Kniffin, Gilbert C. "Streight's Raid Through Tennessee and Northern Georgia in 1863." In *Military Order of the Loyal Legion of the United States*, vol. 4, 193–202.

Kramer, Arnold. *Prisoners of War: A Reference Handbook*. Westport, Conn.: Praeger Security International, 2008.

Krick, Robert K. *Civil War Weather in Virginia*. Tuscaloosa: University of Alabama Press, 2007.

Lewis, John W. "Libby." In *Military Order of the Loyal Legion of the United States*. Vol. 44, 287–304.

"Libby Prison Break." *Civil War History* 4, no. 2 (June 1950): 93–104.

Libby Prison Minstrels. "Programme." Broadside program from the Libby Prison minstrel shows. December 1863. Duke University Special Collections Library.

Libby Prison War Museum Catalogue and Program. Chicago: Libby Prison War Museum Association, 1889. Duke University Special Collections Library.

Libby, Luther, Jr. "Luther Libby: The Man and the Prison." *Richmond Literature and History Quarterly* 1, no. 4 (Spring 1979): 49–50.

Lieber Code, International Committee of the Red Cross. At www.icrc.org.

Lieber, Francis. *The Life and Letters of Francis Lieber*. Edited by Thomas Sergeant Perry. Boston: J. R. Osgood and Company, 1882.

Lincoln, Abraham. *The Collected Works of Abraham Lincoln*. Edited by Roy P. Basler. Vol. 7. New Brunswick, N.J.: Rutgers University Press, 1953.

Lippincott, George E. "Lee-Sawyer Exchange." *Civil War Times Illustrated* 1 (June 1962): 39–40.

Litchfield, A. C. "In the Cell at Libby." In *Prison Life in the South at Richmond, Macon, Savannah, Charleston, Columbia, Charlotte, Raleigh, Goldsborough, and Andersonville During the Years 1864 and 1865*. Edited by A. O. Abbott, 257–259. New York: Harper & Brothers, 1865.

Lloyd, Clive L. *A History of Napoleonic and American Prisoners of War, 1756–1816: Hulk, Depot and Parole*. Woodbridge and Suffolk, U.K.: Antique Collectors' Club, 2007.

Long, E. B., and Barbara Long. *The Civil War Day by Day: An Almanac, 1861–1865*. Garden City, N.Y.: Doubleday & Company, 1971.

MacCauley, Clay. *Through Chancellorsville, into and out of Libby Prison*. Providence: Soldiers and Sailors Historical Society of Rhode Island, 1904. Duke University Special Collections Library.

Magazine of American History 1–30 (1877–1893). New York: A. S. Barnes. Library of Virginia.

Markle, Donald E. *Spies and Spymasters of the Civil War*. New York: Hippocrene Books, 2000.

Martinez, J. Michael. *Life and Death in Civil War Prisons. The Parallel Torments of Corporal John Wesley Minnich, C.S.A., and Sergeant Warren Lee Goss, U.S.A.* Nashville, Tenn.: Rutledge Hill Press, 2004.

McCagg, Ezra B. "The United States Sanitary Commission." In *Military Order of the Loyal Legion of the United States.* Vol. 10, 477–517.

McClure's Magazine 1–44 (1893–1915). New York, London: S. S. McClure Company.

McCreery, William B. *My Experience as a Prisoner of War, and Escape from Libby Prison.* Detroit: Winn & Hammond, 1893. Duke University Special Collections Library.

McGuire, Judith W. *Diary of a Southern Refugee During the War.* Salem, N.H.: Ayer Company, 1986. First published in 1867.

McPherson, James M. *Ordeal by Fire: Volume II, The Civil War.* New York and San Francisco: McGraw-Hill, 1993.

———. *Tried by War: Abraham Lincoln as Commander in Chief.* New York: Penguin Press, 2008.

———. *What They Fought For, 1861–1865.* Baton Rouge and London: Louisiana State University Press, 1994.

McWhiney, Grady, and Perry D. Jamieson. *Attack and Die: Civil War Military Tactics and the Southern Heritage.* Tuscaloosa: University of Alabama Press, 1982.

Military Order of the Loyal Legion of the United States. 66 vols. Wilmington, N.C.: Broadfoot Publishing Company, 1991–1997.

Miller, Francis Trevelyan, ed. *The Photographic History of the Civil War.* 10 vols. New York: Review of Reviews Company, 1911.

Mines, John F. "Life in a Richmond Prison." Excerpted in *Southern Historical Papers* 10, 333–334.

Mitchell, John. "Tunneling out of Libby Prison." *Confederate Veteran* 17 (April 1909): 114. Davis Library, University of North Carolina at Chapel Hill. Microfilm, serials 1–423.

Mitchell, Reid. "Our Prison System, Supposing We Had Any: The Confederate and Union Prison Systems." In Förster and Nagler's *On the Road to Total War,* 564–585.

Moore, Frank, ed. *The Civil War in Song and Story, 1860–1865.* New York: P. F. Collier, 1889.

———. *The Rebellion Record: A Diary of American Events, with Documents, Narratives, Illustrative Incidents, Poetry, Etc.* New York: D. Van Nostrand, 1865.

Moran, Frank E. *Bastiles of the Confederacy*. Baltimore, 1890.

———. "Colonel Rose's Tunnel at Libby Prison." *Century Magazine* 35 (1887–1888): 770–790.

———. "Colonel Rose's Tunnel at Libby Prison." In *Famous Adventures and Prison Escapes of the Civil War*, 184–242.

———. *A Thrilling History of the Famous Underground Tunnel at Libby Prison*. Pamphlet reprint of an 1889 *Century Magazine* article. Duke University Special Collections Library.

Morgan, Michael, "Escape from Rat Hell." *Civil War Times Illustrated* 40 (October 2001): 29–37.

———. "Libby: Warehouse to Big House." *Civil War Times Illustrated* 40 (October 2001): 36–37.

Murray, George W. *Incidents in the Life of George W. Murray During Four Years of Service in the War for the Union, and His Long Confinement and Suffering in Libby Prison*. Cleveland: Fairbanks, Benedict & Company, 1865.

Musicant, Ivan. *Divided Waters: The Naval History of the Civil War*. New York: HarperCollins, 1995.

National Tribune. 1877–1905. Washington, D.C.: G. E. Lemon & Company. Duke University, paper copies, 1877–1889. Library of Congress, Microfilm 1015, reels 2–17, March 1882–December 1905.

The National Tribune Scrap Book: Stories of the Camp, March, Battle, Hospital and Prison, Told by Comrades. 3 vols. Washington, D.C.: National Tribune, 1909.

Neely, Mark E., Jr. "Was the Civil War a Total War?" In Förster and Nagler's *On the Road to Total War*, 29–51.

New York Herald. 1861–1865. New York: James Gordon Bennett. Library of Virginia. Microfilms 1460 and 2330.

New York Times. 1861–1865. New York: H. J. Raymond & Company. Davis Library, University of North Carolina at Chapel Hill. Microfilm, serial 1–517. Archives available online at www.nytimes.com.

Nofi, Albert A. *A Civil War Treasury*. Conshohocken, Penn.: Combined Books, 1992.

Oake, William Royal. *On the Skirmish Line Behind a Friendly Tree*. Helena, Mont.: Farcountry Press, 2006.

Oakley, J. M. "From Libby to Freedom." *Lippincott's Monthly Magazine* 41 (January–June 1888): 812–825.

Obreiter, John. *The Seventy-Seventh Pennsylvania at Shiloh: History of the Regiment*. Harrisburg, Penn.: Harrisburg Publishing Company, 1908.

Parker, David B. *A Chautauqua Boy in '61 and Afterward*. Boston: Smell, Maynard and Company, 1912.

Parker, Sandra V. *Richmond's Civil War Prisons*. Lynchburg, Va.: H. E. Howard, 1990.

Peelle, Margaret W., ed. *Letters from Libby Prison*. New York: Greenwich Book Publishers, 1956.

Petrie, Steward J. *Captive of Libby Prison*. Raleigh, N.C.: Pentland Press, 2002.

Phillips, John W. "Experiences in Libby Prison." In *Military Order of the Loyal Legion of the United States*. Vol. 14, 54–73.

Pierson, J. Frederic. "Prisoners Club, Libby Prison." 1862. Copy of handwritten document. Virginia Historical Society.

Pinkerton, Allan. *The Spy of the Rebellion*. Lincoln: University of Nebraska Press, 1989.

Prutsman, C. M. *A Soldier's Experience in Southern Prisons*. New York: Andrew H. Kellogg, 1901.

Putnam, George Haven. *A Prisoner of War in Virginia, 1864–5*. New York and London: G. P. Putnam's Sons, 1912.

Putnam, Sallie Brock. *Richmond During the War: Four Years of Personal Observation*. Lincoln: University of Nebraska Press, 1996.

Rathborne, St. George. *Fredericksburg: or The Great Tunnel at Libby: A Story of Battle Field and Prison Pen*. New York: Novelist Publishing Company, 1883.

Record of the Federal Dead Buried from Libby, Belle Isle, Danville & Camp Lawton, and at City Point, and in the Field before Petersburg and Richmond. Philadelphia: James B. Rodgers, 1866.

Reynolds, Daniel N. "Memories of Libby Prison." *Michigan History Magazine* 23 (1939): 391–398. Lansing: Michigan Historical Commission.

Richardson, Albert D. *The Secret Service, the Field, the Dungeon, and the Escape*. Hartford, Conn.: American Publishing Company, 1865.

Richmond Daily Whig. 1861–1865. Library of Virginia. Microfilms 144, 144a, 2303, and 2330.

Richmond Dispatch. 1861–1865. Library of Virginia. Microfilms 397, 1934, 2252, and 2331.

Richmond Literature and History Quarterly 1 and 2 (Spring 1979–Spring 1980).

Roach, A. C. *The Prisoner of War, and How Treated*. Indianapolis: Robert Douglass, 1887.

Roe, Alfred S. "Richmond, Annapolis, and Home." In *Personal Narratives of Events in the War of the Rebellion, Being Papers Read Before the Rhode Island*

Soldiers and Sailors Historical Society. Providence: Soldiers and Sailors Historical Society of Rhode Island, 1892. Series 4, no. 10, pamphlet, 1–41.

Rose, Thomas E. *Col. Rose's Story of the Famous Tunnel Escape from Libby Prison.* No publisher listed, 1890s. Duke University Special Collections Library.

———. "Libby Tunnel: An Interesting Account of Its Construction." *National Tribune,* May 14, 1885. Library of Congress, Microfilm 2017.

Ryan, David D. *Cornbread and Maggots—Cloak and Dagger: Union Prisoners and Spies in Civil War Richmond.* Richmond, Va.: Dietz Press, 1994.

———. "Thomas McNiven: Scotsman, Baker, Union Spy." *Civil War* 23, no. 62 (June 1997): 34–37.

Sabre, G. E. *Nineteen Months a Prisoner of War.* New York: American News Company, 1866.

Sanderson, J. M. *My Record in Rebeldom.* New York: W. E. Sibell, 1865.

Schultze, Duane. *The Dahlgren Affair: Terror and Conspiracy in the Civil War.* New York and London: W. W. Norton & Company, 1998.

Schwab, John Christopher. "Prices in the Confederate States, 1861–65." *Political Science Quarterly* 14 (1899): 281–304. New York, Boston, and Chicago: Ginn and Company. Edited by Columbia University political science faculty.

Sclater, W. S. *A Complete and Authentic History of Libby Prison.* Richmond, Va.: Southern Art Emporium, 1897.

Shrady, John. "Reminiscences of Libby Prison." *Magazine of American History* 16 (July 1886): 89–97.

Simpson, Thomas. "My Four Months as a Prisoner of War." In *Personal Narratives of Events in the War of the Rebellion, Being Papers Read Before the Rhode Island Soldiers and Sailors Historical Society.* Providence: Soldiers and Sailors Historical Society of Rhode Island, 1892. Series 3, no. 2, pamphlet, 1–40. Duke University Special Collections Library.

Simson, Jay W. *Naval Strategies of the Civil War.* Nashville, Tenn.: Cumberland House, 2001.

Small, Abner R. *The Road to Richmond: The Civil War Memoirs of Major Abner R. Small of the Sixteenth Maine Volunteers, Together with the Diary Which He Kept When He Was a Prisoner of War.* New York: Fordham University Press, 2000.

Sneden, Robert Knox. Diary 1861–1865. Vol. 5 (December 1863–February 1864), 192–385. Robert Knox Sneden Papers, Virginia Historical Society.

Southern Historical Society Papers. 52 vols. 1876–1959. Richmond, Va.: William Ellis Jones.

Speairs, Arabella, and William Pettit. *Civil War Letters of Arabella Speairs and William Beverley Pettit of Fluvanna County, Virginia, March 1862–March 1865.* Vol. 1. Edited by Charles W. Turner. Roanoke: Virginia Lithography & Graphics Company, 1988.

Speer, Lonnie R. *Portals to Hell: Military Prisons of the Civil War.* Mechanicsburg, Penn.: Stackpole Books, 1997.

———. *War of Vengeance: Acts of Retaliation Against Civil War POWs.* Mechanicsburg, Penn.: Stackpole Books, 2002.

Spencer, William H. *How I Felt in Battle and in Prison.* Privately published. 1860s. Duke University Special Collections Library.

Spies, Scouts and Raiders: Irregular Operations. New York: Time-Life Books, 1985.

Sprague, Homer B. *Lights and Shadows in Confederate Prisons: A Personal Experience, 1864–5.* New York and London: G. P. Putnam's Sons, 1915.

Starkey, Armstrong. *War in the Age of Enlightenment, 1700–1789.* Westport, Conn.: Praeger, 2003.

Stern, Philip Van Doren. *Secret Missions of the Civil War.* New York: Bonanza Books, 1959.

Still, William N., Jr., John M. Taylor, and Norman C. Delaney. *Raiders & Blockaders: The American Civil War Afloat.* Washington, D.C., and London: Brassey's, 1998.

Sturgis, Thomas. *Prisoners of War, 1861–1865: A Record of Personal Experiences, and a Study of the Condition and Treatment of Prisoners on Both Sides During the War of the Rebellion.* New York: Heritage Books, 2003. First published in 1912.

Surdam, David G. *Northern Naval Superiority and the Economics of the American Civil War.* Columbia: University of South Carolina Press, 2001.

Szabad, Emeric. *The Libby Prison Diary of Colonel Emeric Szabad.* Edited by Stephen Beszedits. Toronto: B&L Information Services, 1999.

Thomas, Emory M. *The Confederate State of Richmond: A Biography of the Capital.* Austin and London: University of Texas Press, 1971.

Thompson, Holland. "Escapes from Prison." In Miller's *The Photographic History of the Civil War,* vol. 7, 138–152.

———. "Exchange of Prisoners." In Miller's *The Photographic History of the Civil War,* vol. 7, 98–122.

———. "Life in the Prisons." In Miller's *The Photographic History of the Civil War,* vol. 7, 124–136.

Tiemann, William F. *Prison Life in Dixie.* Typescript. Brooklyn, N.Y., 1894.

Tillson, William H. Diary, 1863–1864. Davis Library, University of North Carolina at Chapel Hill. Microfilm 1–4485.

Tobie, Edward P. *A Trip to Richmond as Prisoner of War*. Providence, R.I.: Sidney S. Rider, 1879.

Tower, Morton. *Army Experience of Major Morton Tower from 1861 to 1864*. Typescript. Virginia Historical Society.

Trumbull, H. Clay. *The Knightly Soldier: A Biography of Major Henry Ward Camp*. Philadelphia: John D. Wattles, 1892.

Turner, Thomas P. Letter to Edith Dabney Tunis Sale. January 6, 1900. Handwritten. Virginia Historical Society.

———. Letter to S. R. Shinn. January 8, 1866. Appeared in the *New York Times* on July 7, 1895, under the headline "Major Turner's Escape." Davis Library, University of North Carolina at Chapel Hill. Microfilm 1–517.

———. Receipts Issued to U.S. Soldiers for Money Confiscated While in Libby Prison. April 27–November 17, 1863. Handwritten. Virginia Historical Society.

———. Special Instructions for the Government of the Guard on Duty at C.S. Military Prisons in the City of Richmond. February 16, 1864. Duke University Special Collections Library.

Turner, William Dandridge. "The Libby Lion." In *The Black Swan* 4 (August 1929): 45, 29–35; (September 1929): 17, 20, 27–29; and (October 1929): 22–23, 33–35. William Turner was Richard Turner's son.

The Union Army: A History of Military Affairs in the Loyal States 1861–65—Records of the Regiments in the Union Army—Cyclopedia of Battles—Memoirs of Commanders and Soldiers. 8 vols. Wilmington, N.C.: Broadfoot Publishing Co., 1997. Reprint of 1908 edition by Federal Publishing.

Urban, John W. *Battle Field and Prison Pen, or Through the War, and Thrice a Prisoner in Rebel Dungeons*. Philadelphia, Boston, and New York: Hubbard Brothers, 1892.

Urquhart, Samuel A. Letter to "My Dear Friend Harris" on Libby Prison Tunnel Association stationery. December 31, 1904. Virginia Historical Society.

U.S. Congress. House. "Mrs. Abby Green." No. 115. *Reports of the Committees of the House of Representatives Made During the First Session of the Thirty-Ninth Congress, 1865–1866*. Vol. 1. Washington, D.C.: Government Printing Office, 1866.

———. "Report on the Treatment of Prisoners of War by the Rebel Authorities During the War of the Rebellion." No. 45. *Reports of Committees of the*

House of Representatives, Third Session of the Fortieth Congress, 1869. Vol. 4. Washington, D.C.: Government Printing Office, 1869.

U.S. Congress. Senate. "Report of the Joint Committee Report on the Conduct of the War." *The Reports of Committees of the Senate of the United States for the Second Session of the Thirty-Eighth Congress.* 3 vols. Washington, D.C.: Government Printing Office, 1865.

———. "Report on Returned Prisoners." No. 68. *Reports of Committees of the Senate of the United States for the First Session of the Thirty-Eighth Congress.* Washington, D.C.: Government Printing Office, 1864.

U.S. Sanitary Commission. *Narrative of Privations and Sufferings of United States Officers and Soldiers While Prisoners of War in the Hands of the Rebel Authorities.* Philadelphia: King & Baird, 1864.

U.S. Senate Executive Documents, 1817–1969. Serial set. Washington, D.C.: Wendell and Van Benthuyson. Davis Library, University of North Carolina at Chapel Hill.

U.S. War Department. *War of the Rebellion: A Compilation of the Official Records of the Union and Confederate Armies.* [Referred to as *O.R.* in Notes.] Series 4, 69 vols. Washington, D.C.: Government Printing Office, 1909. At http://digital.library.cornell.edu/m/moawar/waro/html.

Vance, Jonathan F., ed. *Encyclopedia of Prisoners of War and Internment.* Santa Barbara, Calif., Denver, Colo., and Oxford, U.K.: ABC-CLIO, 2000.

Van Lew, Elizabeth. Album, 1845–1897. Photocopy. Virginia Historical Society.

———. Papers, 1862–1901. Library of Virginia Microfilm, miscellaneous reel no. 14.

———. *A Yankee Spy in Richmond: The Civil War Diary of "Crazy Bet" Van Lew.* Edited by David Ryan. Mechanicsburg, Penn.: Stackpole Books, 1996.

Varon, Elizabeth. *Southern Lady, Yankee Spy: The True Story of Elizabeth Van Lew, A Union Agent in the Heart of the Confederacy.* Oxford and New York: Oxford University Press, 2003.

———. "True to the Flag: Uncovering the Story of Elizabeth Van Lew and Richmond's Union Underground." *North & South* 6, no. 6 (September 2003): 66–81.

Waitt, Robert W. *Libby Prison, Richmond, Virginia.* Richmond, Va.: Richmond Civil War Centennial Committee, 1962.

Wallber, Albert. "The Escape from Libby." In Domschke's *Twenty Months in Captivity,* 125–137.

Walls That Talk: A Transcript of the Names, Initials and Sentiments Written and Graven on the Walls, Doors and Windows of the Libby Prison at Richmond.

Richmond, Va.: R. E. Lee Camp, No. 1, C.V., 1884. Duke University Special Collections Library.

Wells, James M. "The Tunnel Escape of Union Prisoners from the Confederate Libby Prison." In *The Most Incredible Prison Escape of the Civil War*, edited by W. Fred Conway, 91–113. New Albany, Ind.: FBH Publishers, 1991.

———. "Tunneling Out of Libby Prison: A Michigan Lieutenant's Account of His Own Imprisonment and Daring Escape." *McClure's Magazine* 22 (November 1903–April 1904): 317–326.

Whitman, Walt. *Complete Poetry and Selected Prose*. Edited by James E. Miller Jr. Boston: Houghton Mifflin Company, 1959.

Wilkins, William D. "Forgotten in the Black Hole: A Diary from Libby Prison." *Civil War Times Illustrated* 15 (June 1976): 36–44.

Willett, Robert L., Jr. "Again in Chains: Black Soldiers Suffering in Captivity." *Civil War Times Illustrated* 20 (May 1981): 36–43.

———. *The Lightning Mule Brigade: Abel Streight's 1863 Raid into Alabama*. Carmel, Ind.: Guild Press, 1999.

Wright, Mike. *City under Siege: Richmond in the Civil War*. Lanham, Md., New York, and London: Madison Books, 1995.

Wyman, Lillie Buffum Chace. *A Grand Army Man of Rhode Island*. Newton, Mass.: Graphic Press, 1925. Duke University Special Collections Library.

Young, Bennett. "Treatment of Prisoners of War." *Confederate Veteran* 26 (November 1918): 470, 501. Davis Library, University of North Carolina at Chapel Hill, Microfilm 1–423.

INDEX

Page numbers listed as ins. *indicate photograph insert*

Abraham, Tom, 141–142
Adams, John, 84
Adler, Adolphus, 90
Alexander, George, 194
American Colonization Society, 85
Anderson, Joseph, 2–3
Andersonville, 133, 210, 213–215
Antietam, battle of, 14–15
Arlington National Cemetery, 223
Atlantic Monthly (magazine), 57
Autobiography (Butler), 213

Baker, Eliza Louise, 84, 85, 86, 88
Baker, Hilary, 84
Bartleson, Frederick, 31, 34, 53, 159,
    163–164, 189
Bartleson, Kate, 34
"Battle Hymn of the Republic, The"
    (Howe), 57, 58
Beaudry, Louis N., 37, 38, 40, 62–63
Belle Isle
    abuses at, 195
    after war, 225–226
    condition of prisoners, 74–75
    Confederate position on conditions at,
       79
    deteriorating conditions at, 197
    Dow and, 69–70
    increase in prisoner deaths, 133
    number of prisoners, 30
    removal of prisoners to Andersonville,
       213
Bellevue Elementary School, 228

Bennett, F. F., 115, 117
Black market, Libby Prison, 53–54, 67
Blacks
    recruitment by Union Army, 28
    Unionist underground movement in
       Richmond and, 88
Black Union war prisoners
    "Great Escape Plot" and, 123
    at Libby, 35, 45, 90, 109, 139, 209
    release of, as condition for prisoner
       exchanges, 81, 211–212, 216,
       217
    Unionists and, 89–90
Bonaparte, Napoleon, 19
Booth, Benjamin, 66, 136
Boston Daily Advertiser (newspaper), 76
Botts, John Minor, 91
Boutelle, Charles, 96
Bowser, William, 100
Boyle, William, 141–142
Bragg, Braxton, 104, 105, 208
Bremer, Fredrika, 84
Brown, John, 201
Brown, Joseph T., 59
Brown, Spencer Kellogg, 97
Browne, Junius H., 35, 55, 60, 61
Bryan, E. Pliny, 79
Buchanan, James, 24
Burnside, Ambrose, 58
Butchertown Cats, 12
Butler, Benjamin, 83, 169, 195, 203, 220,
    ins. pg. 3
    box deliveries and, 138

Butler, Benjamin (*continued*)
  prisoner exchanges and, 80–81, 211,
    212, 213
  raid to free Union prisoners, 129–131,
    140–142
  reward to free black man for aid to
    Union prisoners, 180
  Southern hatred of, 80
  on treatment of Dahlgren's body,
    207
  Van Lew and, 96–99, 226
Byers, H. M., 68
Byers, S. H. M., 141

Caffey, John, 180
Caldwell, David
  on escape preparations, 121, 126, 152,
    165, 171
  following escape, 178–179
  on Libby's July 4th celebration, 56
  reflection on confinement, 44
  on scaling process, 51
Camp Chase (Ohio), 27
Camp Douglas (Illinois), 25, 27
Camp Fires and Camp Cooking
  (Sanderson), 63
Camp Morton (Indiana), 26
Camp Sumter (Andersonville), 133, 210,
  213–215
Carrington, Isaac, 78–79
Carrington, William A., 133
Cartel. *See* Prisoner exchanges
Carson, John A., 26
Cashmeyer, Philip, 97
Casualties, Confederate, in Richmond,
  4–5, 7–8
Cavada, Federico
  on captivity, 138
  on deprivations of Southerners,
    139–140
  on escape, 103
  on escape preparations, 111
  on Lyceum, 37
  on physical surroundings at Libby, 54
  on roll call, 193
  on Sanderson, 63
  on shortage of rations, 136
  Spanish classes at Libby and, 39
Chamberlain, J. W., 26, 53–54, 59
Chandler, D. T., 214

Charleston Daily Courier (newspaper),
  12–13
Chase, Samuel P., 99
Chesnut, Mary, 13, 15, 16, 60, 134, 210
Chicago, Libby Prison War Museum,
  224, ins. pg. 9
Chicago Coliseum, 225
Chicago Historical Society, 225
Chickahominy River, 177–178, 181, 185,
  189
Chickamauga, battle of, 104–105
Chimborazo Hospital, 8, 16
Christian faith, Libby Prison captives
  and, 44
City Point (Virginia), ix–x, 42, 72, 73, 97,
  98, 195, 223, ins. pg. 3
Clarke, James Freeman, 57
Classes, at Libby Prison, 36–37
Cobb, Howell, 21, 22
"Cockfights," at Libby Prison, 41–42
*Code for the Government of Armies in the
  Field* (Union Army), 20–21
Code key, Van Lew, 98, 228, ins. pg. 5
Commandants, Libby Prison, 51–52
Confederate casualties in Richmond, 4–5,
  7–8
Confederate government
  on black prisoners of war, 28, 29
  breakdown of prisoner exchange and,
    28, 29–30
  food riot and, 9–10
  plan to move Union prisoners into
    Deep South, 130–133, 213
  prisoner food boxes and, 68, 73, 81–82
  public relations regarding war prisoner
    abuses, 78–79
  reaction to Dahlgren's address to his
    men, 206–207
  response to Libby Prison breakout,
    173–176
  war prisons report, 216
Confederate guards, 37, 41, 44, 93, 137
  food boxes and, 68, 82
  increased security measures, 71, 126,
    193–194
  investigating during tunnel excavation,
    145, 146, 149–151, 158
  Libby Prison escape and, 108,
    110–111, 143, 163–164, 167–168,
    173–174

McCullough escape and, 95
shopping for prisoners, 33, 34, 53, 54, 67
Streight and, 62
Van Lew and, 87
Confederate parolees, 29, 77, 211, 212, 217
Confederate View of the Treatment of Prisoners (Jones), 217
Confederate war prisoners, 26–27, 77
Confederate war prisons, 24–25
    Andersonville, 133, 210, 213–215
    See also Belle Isle; Libby Prison
Corbin, Melissa, 58
Corbin, William Francis, 58–59
Cornwell, Lydia, 67
Cornwell, Robert T., 37, 67, 137, 157, 223
Counterfeiting, of greenbacks, 54
Custer, George Armstrong, 199

Dahlgren, John A., 200, 202, 207, 208
Dahlgren, Ulric, 193, 199, 200–202, 205–208
Daily Enquirer and Examiner (newspaper), 226–227
Daily Richmond Enquirer (newspaper), 174, 175, 176, 182, 185
Daily Richmond Examiner (newspaper), 141
    Bott's letter to, 91
    on Butler, 81
    on Dahlgren burial, 207
    on Dog War, 12
    on food riot, 10
    on "Great Escape Plot," 123
    on prisoner exchange, 79
Daily routine, at Libby Prison, 34–35
Darby, George, 52
Davidson, H. M., 66
Davis, George, 178
Davis, Jefferson, 3, 29, 93, 131
    arson attempt on house of, 134
    on Butler, 80
    on Confederate parolees, 211
    Dahlgren orders and, 206, 207
    Emancipation Proclamation and, 27–28
    exhorting Southern populace, 61
    food riot and, 9

"Great Escape Plot" and, 123
    indicted for war crimes, 215
    Jackson funeral procession and, 14
    martial law in Richmond and, 5
    on mistreatment of prisoners, 217
    rock fights and, 12
    rooting out Union sympathizers, 90–91
Davis, Varina Howell, 206–207
Deaths. *See* Mortality
Deaton, Spencer, 194–195
DeLeon, T. C., 1, 3, 5, 14
Di Cesnola, Louis, 39
Disease
    among Rebel prisoners, 27
    at Belle Isle, 30
    at Libby Prison, 30, 62
Dix, John A., 22, 23
Dog(s)
    Libby Prison, 194
    search for escapees and, 178, 186
Dog War, 12
Domschke, Bernhard, 36, 52, 140
Dougherty, Michael, 66
Douglass, Frederick, 29
Dow, Neal, 68
    "Great Escape Plot" and, 123
    head count and, 172
    leading "double-quick," 137
    Libby Prison escape and, 159
    Morgan and, 136
    postwar career, 223–224
    report on conditions at Belle Isle, 69–71, 72, 74
    Sanderson and, 198
Driscoll, Edward, 55
Dungeon, at Libby Prison, 44–46

Earle, Charles, 159, 162, 170
Early, Jubal, 221
Elite of Richmond, life during war, 13–14
Elmira prison, death rate, 27
Ely, Alfred, 37–38, 87
Ely, William, 123
Emancipation Proclamation, 15–16, 80
    effect on prisoner exchanges, xi–xii, 27–28
Encyclopedia Americana, 19
Enders, John, 31
Engles, Caroline Cox, 6

Executions
  Deaton, 194–195
  proposed, of Dahlgren's men,
    208–209
  proposed, of Flynn and Sawyer, 59–60
  Webster, 91, 97
  Wirz, 215

Farragut, David, 80
Fischer, David Hackett, 22
Fislar, J. C., 148, 170, 179–180
Fitzsimmons, George, 114
Flynn, John P., 59–60
Foley, Joseph, 53
Food riot, 8–10
Food shortages
  Confederate, 75, 215
  Libby Prison, 137
  in Richmond, 8–11, 133–134, 139–140
Ford, Robert, 89, 101, 183, 188
Forrest, Nathan Bedford, 48, 49, 50
Forsyth, George, 71
Fortress Monroe, ix, 129–131, 140
Foundries, Richmond, 2
Fox, Edward, 205
Francis, Lewis, 87
Free-food depots, 10
Fugitives, Richmond Unionists and,
    89–90, 183–184

Gallagher, John, 150
Games, played at Libby Prison, 35, 41
Gangs, Richmond, 12
Garfield, James A., 48
Gates, Junius, 187
General Order No. 28, 80
General Order No. 38, 58
General Order No. 100, 20–21
General Order No. 252, 29
Geneva Conventions, 20
Gettysburg, battle of, 57, 58
Glazier, Willard, 35, 36, 74, 137, 165–166
Goss, Warren Lee, 138–139
Graffiti, Libby Prison, 36
Grant, George, 43
Grant, Julia, 221
Grant, Mary, 43
Grant, Ulysses S., 29, 60, 61, 78, 90, 98,
    215
  death mask, 224

prisoner exchanges and, 211–213,
    217–218
  Van Lew and, 220, 221, 226
Gray, Charles Carroll, 33
"Great Escape Plan, The," 123
"Great Yankee Wonder," 174
Green, Abby, 88, 89, 101, 123, 183
Greenbacks, 52–54
Gunther, Charles F., 224

Hairston, J. T. W., 51–52
Hall, James O., 206
Hallbach, Edward, 205
Halleck, Henry "Old Brains," 20, 76
Hamilton, A. G., ins. pg. 2
  final escape, 171, 189
  first escape attempt, 107–109
  first tunnel excavation, 111–120
  fourth tunnel excavation, 143–145,
    151, 154–155
  second escape attempt, 109–111
  second tunnel excavation, 121–122
  third tunnel excavation, 125, 128
Hamlin, Hannibal, 57
Hammond, Daniel, 72
Hampton, Wade, 199, 203, 209
Harper's Weekly (newspaper), 74, 180
Hayes, Rutherford, 227
Head counts at Libby Prison, 41, 93–94,
    125–127, 145, 147, 172, 193
Heffley, Albert, 193
Heffley, Cyrus P., 82, 129, 138, 204,
    213
Henry, John, 39
Hesseltine, William, 24
Hickey, John, 72
Hill, A. P., 135
Hill, D. H., 22, 23
Hitchcock, Ethan Allen, 30, 77, 79, 195
Hobart, Harrison C.
  escape, 170, 178
  as gatekeeper, 160, 161, 162, 164
  Libby Prison Tunnel Association and,
    224
  prisoner exchange and, 195
Hoffman, William, 25, 26
Holmes, Arnold B., 88, 95
Holmes, Josephine, 95
Homes of the New World (Bremer), 84
Housum, Peter, 104

Howard, Harry S., 95
Howe, Julia Ward, 57
Hurlbert, William Henry, 90
Huson, Calvin, 87
Hussey, John, 70

Independence Day celebration at Libby
  Prison, 55–57
Independent Provisional Brigade, 47,
  49–50
Intelligence gathering
  among black prisoners and black
    servants, 46
  by Unionists, 94–99, 100
  by Van Lew, 94–95, 96–99, 100
Invisible ink messages, 43–44, 96–97, 98
Iron making, in Richmond, 2–3
Isham, Asa, 214

Jackson, Thomas "Stonewall," 14
James River Park System, 226
Jews in Richmond, 4
  scapegoating of, 134–135
Johnston, Albert Sidney, 15
Johnston, I. N., 66
  confinement to east cellar, 147,
    148–150
  escape, 158, 163, 170, 179–180
  on escape attempts, 121, 128
  tunnel excavation, 145, 146, 153, 154,
    176
Johnston, Joseph, 211, 221
Joint Select Committee on the Conduct
  of the War, 197, 210
Jones, Eliza A., 129
Jones, James Ap., 97, 98, 129
Jones, John B., 16, 61, 134
Jones, Virgil Carrington, 206
Jos. R. Anderson & Co., 2

Keeler, Alonzo, 42–43
Kellogg, F. W., 196
Kendrick, W. P., 180, 195
Kentucky, 58–59
Key, Francis Scott, 23
Key, Philip Barton, 23–24
Kilpatrick, Hugh Judson, 199–205

Latouche, John, 93, 137, 147, 173–174,
  188

Lawton, Hattie, 91
Leaves of Grass (Whitman), 196
Lee, Robert E., 57, 68, 104, 199, 206, 211
  on executing Dahlgren's men, 208–209
  prisoner exchange and, 22
  on removing Union prisoners south,
    132–133
  statue of, 227
  surrender of, 221
Lee, W. H. F. "Rooney," 59–60, 77
Letcher, John, 8, 9
Letters, Yankee war prisoner, 138, 194,
  ins. pg. 5
Lewis, John, 167–168, 187
Lewis, Oliver, 88
Libby, George, 32
Libby, Luther, 32
Libby Chronicle (newspaper), 37, 38–40,
  62–63, 65
Libby Prison
  arrival of new prisoners, 54–55
  black market, 53–54, 67
  Christian faith and, 44
  commandants, 51–52
  Confederate spies, 117
  daily routine, 34–35
  description of, 32–34
  deterioration of conditions in, 136–139
  dungeon, 44–46, 187–188, 191, 222
  escape from, 93–94, 95–96, 126,
    157–168
  "Great Escape Plot," 123–125
  increased security measures, 193–194
  Independence Day celebration, 55–57
  mail calls, 42–44
  number of prisoners, 30, 33
  officer escape January 1864, 126
  as officer prison, 25
  order against standing near window,
    71–72
  practical jokes among prisoners, 40–42
  prisoner activities, 35–37
  prisoner burials, 139
  prisoner food boxes, 67–68, 73, 81–82
  prisoner organizations, 37–38
  prison newspaper, 37, 38–40, 62–63, 65
  public knowledge of conditions at,
    62–64, 73–74
  raid on Richmond and, 203–205
  rations, 33, 62, 63, 65–67, 137

Libby Prison (*continued*)
  rats, 45, 107, 116, 118
  Rebel plan to avenge deaths of Corbin
    and McGraw, 58–60
  representation of, ins. pg. 1
  reversion to warehouse after war, 224
  scaling process, 51, 52–53
  slave laborers, 45–46
  Streight at, 54–55, 56, 61–64
  tourists to, 135
Libby Prison Association, 55, 56
Libby Prisoners Club, 38
Libby Prison escapees, 94, 95–96, 126,
    160–168, 170–191
  enumeration of abuses by, 195–198
  re-creation of capture of, ins. pg. 7
Libby Prison escape tunnel, ins. pg. 6, ins.
    pg. 7. *See also* Hamilton, A. G.;
    Rose, Thomas Ellwood
Libby Prison Tunnel Association, 224
Libby Prison War Museum, 224–225, ins.
    pg. 8
Libby Prison War Museum Corporation,
    224
Lice, 35
Lieber, Francis, 19–21
Lieber Code, 20–21, 28
Lightning Mule Brigade, 49–50, 61
Lincoln, Abraham, 57, 158, 195
  appeal to stop Sawyer execution, 59
  assassination attempt, 91
  bust at Libby Prison, 36
  clemency appeal, 58
  Emancipation Proclamation, 15–16,
    27
  Grant and, 211, 217–218
  prisoner exchanges and, 21
  raid on Richmond and, 198–199, 200
  reaction to idea of retributive
    measures, 76
  suspension of military executions,
    141
  suspension of prisoner exchanges, 29
  tour of Richmond, 221
Lincoln, Tad, 221
Lind, Jenny, 84
Lipscomb, Martin, 97
Litchfield, A. C., 209
Littlepage, William, 205
Lockwood, Henry Hayes, 131

Lohmann, F. W. E., 88
Longstreet, James, 104
Lownsbury, William, 93–94
Lyceum (Libby Prison debating society),
    37

MacCauley, Clay, xi
Mail, Confederate censorship of, 43–44
Mail calls, at Libby Prison, 42–44
Manassas, battle of, 4–5, 7
Marshall, John, 84
Mayo, Joseph, 5, 9, 219
McCabe, Charles, 36–37, 39, 44, 53, 57,
    139
McCabe, Kate, 37
McClellan, George, 7, 22, 92, 131
McCreery, William, 195
McCullough, John R., 95–96
McDonald, B. B.
  escape, 183, 184, 185, 187
  missing at head count, 147–148
  Sanderson and, 159
  tunnel excavation and, 118, 152–154
  Van Lew and, 184, 225
McGill, John, 44
McGraw, Thomas Jefferson, 58–59
McGuire, Judith, 3, 10, 12, 14, 15
McKean, N. S., 170
McKean, Nineoch, 181
Meade, George, 130, 199, 206
Memminger, Christopher, 86
Meredith, S. A., 62, 68, 70, 72–73, 81
Meredith, Sam, 124
Milroy, Robert, 29
Moore, S. P., 7–8
Moot court, Libby Prison, 36
Moran, Frank
  escape, 164–165, 166
  on guards shooting prisoners at
    windows, 71
  recapture of, 175–176
  on Rose, 106, 118, 143, 152, 154
  on tunnel excavation, 125, 127, 143,
    152, 154
  Turner and, 51
Morgan, Charles, 166, 175
Morgan, John, 185
Morgan, John H., 135–136
Mortality
  at Andersonville, 214, 215

at Confederate war prisons, 27, 73, 214, 215, 216–217
at Richmond war prisons, 73
at Union war prisons, 27, 216–217
Musgrave, Jonathan, 26
Music, at Independence Day celebration, 56–57
Myers, A. C., 24
Myers, W. W., 72
My Record in Rebeldom (Sanderson), 198

National Intelligencer (newspaper), 59, 74, 78, 123, 176, 185, 187
"Nero," 195
New Orleans, capture of, 15, 80
Newspaper reports of Libby Prison escape, 174–175, 176
New York Herald (newspaper), 132, 196
New York Times (newspaper), 29–30, 35, 74, 76, 125, 197, 212
New York Tribune (newspaper), 55

Ohio State Penitentiary, Morgan escape from, 135–136
Old Dominion Iron and Steel Co., 225
Ould, Robert
Dahlgren burial and, 207
Dow at Belle Isle and, 70
food boxes and, 81–82, 138
meeting with Meredith, 72–73
prisoner exchanges and, 19, 23–24, 29, 30, 61, 62, 142, 211, 212
on providing for Union prisoners, xi
Vicksburg parolees and, 29

Palmer, Charles, 130
Parker, David, 90, 220–221
Pemberton, John, 61
Philadelphia Bulletin (newspaper), 74
Pickett, George, 123
Pinkerton, Allan, 91
Pleasanton, Alfred, 199
Plug-uglies, 6–7, 91, 188
Poe, Edgar Allen, 84
Point Lookout (Maryland), 25–26
Pollard, James, 205
Population growth, of Richmond, 3–4
Porter, John F., 126
Powell, William, 123, 125

Practical jokes, among Libby Prison captives, 40–42
Prisoner exchanges, ix–xii
Butler and, 80–81
calls for renewed, 76
Civil War, 21–23, 27–30
Confederate request for, 215–216
Grant and, 211–213, 217–218
Lieber Code and, 20, 21
Ould and, 19, 23–24, 29, 30, 61, 62, 142, 211, 212
suspension of, 27–30
Union insistence Rebels return to inactive service, 77–78
Prisoner food boxes, 67–68, 73, 81–82, 138
Prisoners of war
black (Union) (*see* Black Union war prisoners)
Confederate, 26–27, 77
Lieber Code and, 20
Revolutionary War, 22
Union (*see* Union war prisoners)
Provost guards (plug-uglies), 6–7, 91, 188
Pryor, Sara A., 8
Public knowledge of conditions at Libby Prison, 62–64, 73–74
Putnam, George, 139, 225
Putnam, Sally, 1, 3, 4, 12, 16, 60, 134, 203

Quarles, John H., 88, 183, 184

Radcliffe, S. J., 74–75
Randall, W. S. B., 146, 151–152, 170–171, 181–182
Randolph, R. W., 159
Rathborne, St. George, 45
Rations
Libby Prison, 33, 62, 63, 65–67, 137
Rebel prisoner, 26–27, 77
Rats, Libby Prison, 45, 107, 116, 118
Reed, B. C. G., 122–123, 124
Refugees, in Richmond, 3
Relief of Poor Persons not in the Poor House, 10
"Repeating," 126–127, 145, 147
Revolutionary War, treatment of prisoners of war, 22
Rice, Lucy, 183–184, 185
Richards, Mary Jane, 85, 88, 100

Richmond
  anarchy in, 12–13
  burning of, 219–220
  as capital of Confederacy, 1, 3–7
  civilian unrest in, 134–135
  Confederate casualties in, 4–5, 7–8
  deterioration during war, 16
  food shortages in, 8–11, 133–134,
    139–140
  hatred of Yankees in, 16–17
  infrastructure, 2
  Libby Prison escapees in, 169–172
  martial law in, 5–7, 91
  population growth of, 3–4
  reaction to Union victory at Vicksburg,
    60–61
  response to Union raid, 203–204
  Union Army threat to, 15
  Unionist underground movement in,
    88–89
  Union raids on, 131, 140–142, 198–205
  women and war effort in, 11–12
Richmond Bread Riot, 9–10
Richmond Daily Whig (newspaper), 13,
  206, 207
Richmond Dispatch (newspaper), 79, 92,
  174, 187, 209
Richmond Evening Journal (newspaper),
  227–228
Richmond Prison Association, 37–38
Richmond Sentinel (newspaper), 185
Roach, A. C., 45, 139, 165, 167, 173
Roane, James and Peter, 88
Rock fights, 12
Rose, Thomas Ellwood, 103–104, 105,
  209, ins. pg. 2, ins. pg. 6
  escape, 157, 161, 163
  escape tunnel plans, 111–112
  first escape attempt, 106–109
  first escape tunnel excavation, 108,
    111–120
  fourth tunnel excavation, 143–155
  "Great Escape Plot," 123, 124
  liberation of, 223
  plan on using sewer for escape, 121–122
  recapture of, 189–191
  second escape attempt, 109–111
  second escape tunnel excavation,
    121–122
  third tunnel excavation, 125–128

Rosecrans, William, 104, 105
Ross, Erasmus, 41, 93–94, 125–127, 145,
  147, 172, 221
Round Mountain Iron Works, 50
Rowan, Charles, 162
Rowley, William, 88, 95, 99, 130, 208
Ruth, Samuel, 88, 99

Sabre, G. E., 70
Sanderson, James, 63, 66
  exoneration of, 198
  "Great Escape Plot" and, 124–125
  McDonald escape and, 159
  Streight and, 63–64, 197–198
Sawyer, Henry Washington, 59–60
Scaling process, 51, 52–53
Scearce, William, 183, 185, 187
Scobell, John, 91
Seddon, James, 71, 201
  Dahlgren's raiders and, 208
  food riot and, 9
  Lee letter to, 132
  mining Libby with explosives and, 204
  on prisoner exchange breakdown, 29
  on raid on Richmond, 209–210
  Streight letter to, 62
  on Union prisoner rations, 137
Seddon, Sarah Bruce, 201–202, 209
Seven Days battles, 7
Sewell, William, 89
Seymour, Truman, 76
Shaffer, J. W., 177, 184–185
Sharpe, George, 226
Sharpe, George Henry, 98
Sherman, William, 65, 78, 215, 223
Shiloh, battle of, 15, 22
Shockoe Hill Cats, 12
Sickles, Daniel, 23–24
Sickles, Teresa, 24
Slave insurrection, fear of, 28, 55, 134
Slaves
  aid to Libby fugitives, 179–180, 186
  industrial, 2–3
  as laborers at Libby Prison, 45–46
  in South after Emancipation
    Proclamation, 15–16
Small, Melville, 176
Sneden, Robert, 169, 174
Southern Fertilizing Company, 224
Southern Punch (magazine), 134

Spalding, A. G., 224
Spies
  Confederate, 58, 117
  hanging of Union, 91
  See also Van Lew, Elizabeth
Stanton, Edwin, 22, 24, 26
  appeal to stop Sawyer execution, 59
  Dow letter to, 70–71, 72
  raid on Richmond and, 129, 130, 131
  reduced prisoner rations and, 77
  resumption of prisoner exchanges and, 81
  suspension of prisoner exchanges, xii, 28
  Union prisoner abuses, 70–71, 72, 195, 210
Stearns, Franklin, 88, 94
Sterling, John, 183, 185, 187
Stoneman, George, 15
Streight, Abel D., 47–51, ins. pg. 1
  arrival at Libby Prison, 54–55
  escape, 159, 167, 171
  flight from Richmond, 185–187
  "Great Escape Plot," 122–125
  impact at Libby Prison, 56, 61–64
  postwar career, 223–224
  reports of escape, 175
  search for fugitive, 177, 182–185
  in Washington after escape, 196–198
Sturgis, Thomas, 26
Sugar shortage, 61
Szabad, Emeric, 82

Thomas, George, 104
Thompson, Jacob, 53
Tobacco trade, Richmond and, 2
Tobie, Edward, ix–x
Todd, David, 52, 86
Total war, 78, 217–218
Tower, Morton, 178, 195–196
Tredegar Iron Works, 2–3, 123, 131, 200
Trumbell, H. C., 70
Trumbower, Lydia C., 104
Turner, Richard, 89, 183, 188
  abuse of prisoners, 51, 52
  Libby Prison escape and, 173, 175–176, 187
  life after surrender, 222
  overcharging of prisoners, 67

Turner, Thomas, 51, 56, 59, 70, 188, ins. pg. 5
  death of, 222
  departure from Libby Prison, 219
  on guards shooting at prisoners, 72
  increase in security measures, 126, 193
  instruction to guards to cease buying food for prisoners, 82
  kicking a dead prisoner, 137
  limiting letters sent by prisoners, 138
  McDonald and, 148
  mining of Libby Prison and, 204
  postwar career, 221–222
  response to escape, 173
  Streight letter to, 62
Turner, William, 222

Union
  demands for retaliation over treatment of Yankee war prisoners, 75–77
  outrage over prisoner abuses in South, 197
Union Army, recruitment of blacks, 28
Union government, prisoner exchanges and, 23, 27–30, 77–78, 80–81, 211–213, 217–218
Unionist underground movement, 88–89, 221
  attempts to root out, 90–92
  Dahlgren's burial and, 208
  helping prisoners escape, 93–97
  intelligence gathering, 94–99
Union war prisoners
  Butler's raid to free, 129, 131, 140–142
  Confederate plan to move into Deep South, 130–133, 213
  demands for retaliation over treatment of Yankee, 75–77
  Kilpatrick's raid to free, 199–205
  reports on treatment of, 70–71, 72, 196–197, 210, 216
  segregation of, 25
  See also Belle Isle; Libby Prison
Union war prisons, 25–26
  punitive measures in, 77
U.S. Christian Commission, 70
U.S. Sanitary Commission, 25, 67, 76, 82, 197, 210
Urban, John, 51

Van Lew, Annie, 83, 96, 220, 228
Van Lew, Eliza, 92, 208
Van Lew, Elizabeth, 134, 185, 219, ins.
    pg. 4
    abetting fugitives, 83, 90, 94, 101, 159
    as abolitionist, 85
    on burning of Richmond, 220
    Butler's raid and, 129–130, 131
    code key, 98, 228, ins. pg. 5
    Confederate scrutiny of, 99–100
    Dahlgren's burial and, 208
    defiance of Confederates, 92–93
    early life, 84–85
    "Great Escape Plot" and, 123
    headstone, 229, ins. pg. 8
    intelligence gathering, 94–95, 96–99,
        100, 198, 213
    Libby fugitives and, 169, 171–172,
        183–184, 225
    as message smuggler, 87
    postwar life and career, 226–228
    protection of home by Union Army,
        220–221
    report on plans for Union prisoners,
        198, 213
    underground movement and, 88–89
    visits to war prisoners, 86–88, 89–90
Van Lew, John, 84, 85
Van Lew, John (Jr.), 88, 100, 171–172
Van Lew mansion, ins. pg. 4
Varon, Elizabeth, 85, 90, 92
Vicksburg, surrender of, 60–61, 211
Vicksburg parolees, 29, 77, 212, 217
Virginia Club, 228
Virginia Electric and Power Co., 225
von Blücher, Gebhard Leberecht, 19

Wallace, Lew, 23
Wallber, Albert, 178, 180
War of 1812, cartel model, 21–22
Washington, George (free black), 180

Wasson, J. M., 170, 175
Watson, William, 166, 175
Webster, Timothy, 91, 97
Wells, James, 157, 164, 166, 181
West, T. S., 164, 170, 178
White, Harry, 110
Whitman, Walt, 196, 212
Wilcox, Harry, 165, 166
Wilkins, William, 35
Willett, Robert, Jr., 50
Williams, Leander, 162
Williams, W. A., 171, 178–179
Winder, John, 23, 32, 61, 70, 91
    in charge of prisons at Macon and
        Andersonville, 214–215
    Libby Prison breakout and, 173, 174
    prisoner transfer to Deep South and,
        133
    as provost marshal, 5–7
    response to Union raid on Richmond,
        203–204
    security at Libby Prison and, 71, 77
    sentence of Ford, 188
    Van Lew and, 86, 172
Winder, Richard, 133, 222
Winder, William Henry, 5–6
Winder, William Sidney, 59–60, 133
Wirz, Henry, 52, 215, 221–222
Wise, Henry A., 201
Wistar, Isaac J., 131, 140–141, 177, 184–
    185, 198
Withers, John, 9
"Woman order," 80
Women
    food riots and, 8–9
    war effort and, 11–12
Wool, John, 21, 22
"Wound-Dresser, The" (Whitman), 196

Yankees, Southern hatred of, 16–17. *See
    also under* Union

PAT WHEELAN

Joseph Wheelan, a former Associated Press reporter and editor, is the author of *Mr. Adams's Last Crusade*, *Invading Mexico*, *Jefferson's War*, and *Jefferson's Vendetta*. He lives in Cary, North Carolina.

PublicAffairs is a publishing house founded in 1997. It is a tribute to the standards, values, and flair of three persons who have served as mentors to countless reporters, writers, editors, and book people of all kinds, including me.

I.F. STONE, proprietor of *I. F. Stone's Weekly*, combined a commitment to the First Amendment with entrepreneurial zeal and reporting skill and became one of the great independent journalists in American history. At the age of eighty, Izzy published *The Trial of Socrates*, which was a national bestseller. He wrote the book after he taught himself ancient Greek.

BENJAMIN C. BRADLEE was for nearly thirty years the charismatic editorial leader of *The Washington Post*. It was Ben who gave the *Post* the range and courage to pursue such historic issues as Watergate. He supported his reporters with a tenacity that made them fearless and it is no accident that so many became authors of influential, best-selling books.

ROBERT L. BERNSTEIN, the chief executive of Random House for more than a quarter century, guided one of the nation's premier publishing houses. Bob was personally responsible for many books of political dissent and argument that challenged tyranny around the globe. He is also the founder and longtime chair of Human Rights Watch, one of the most respected human rights organizations in the world.

·   ·   ·

For fifty years, the banner of Public Affairs Press was carried by its owner Morris B. Schnapper, who published Gandhi, Nasser, Toynbee, Truman, and about 1,500 other authors. In 1983, Schnapper was described by *The Washington Post* as "a redoubtable gadfly." His legacy will endure in the books to come.

Peter Osnos, *Founder and Editor-at-Large*